THE PIGMAN'S HANDBOOK

THE PIGMAN'S HANDBOOK

GERRY BRENT

FARMING PRESS LIMITED

First published 1982

Second Edition 1987

Reprinted 1989 & 1995

Copyright © FARMING PRESS LTD, 1982, 1983, 1987

Farming Press Books & Videos
Wharfedale Road, Ipswich IP1 4LG
Suffolk, United Kingdom

Distributed in North America by
Diamond Farm Enterprises
Box 537, Alexandria Bay, NY 13607, USA

Brent, Gerry
The Pigman's Handbook. – 2nd ed.
1. Swine
I. Title
636.4   SF395
ISBN 0-85236-170-X

Printed in Great Britain by Butler & Tanner Ltd
Frome and London

CONTENTS

ILLUSTRATIONS

PREFACE

WHEN THE original text of *The Pigman's Handbook* was prepared in early 1982 there was a feeling that little would change in the timeless application of the stockman's skills to justify a revision within five years. This was particularly believed to be so when the book is substantially a compendium of common sense and of logically reviewing husbandry practices.

The fact that a revision has been produced is an indication that experience has shown that emphasis has shifted in the quest for greater efficiency. Either that or the author has been made painfully aware that some aspects of the original text did not place sufficient accent on particular matters! Whatever the interpretation of the need, the modest aim of *The Pigman's Handbook* remains unaltered. It is intended as a 'prop' in times of seemingly confused and obstinate production problems to ensure that the lonely pig operator is giving due consideration to all those factors which might assist in a resolution of his difficulties.

Increasingly, people contend that the importance of skilled pigmen – and some of the best pigmen I know are women– is on the wane due to greater application of technology. Against a background of declining profits per pig produced, the author cannot visualise the time when a proprietor will not seek to locate, retain and reward those who can achieve greater fiscal returns on his behalf.

It may well be that the pigman of the future will have to become more conscious of the need to tailor results to financial rather than the more accepted physical results. The key to doing just that is to be aware of production output so that manipulation can be made. In any case the pigman needs, above most else, to be able to produce pigs despite the disadvantages of buildings, diet and pig health that may confront him. Future revisions may well need to place still more emphasis upon record interpretation as a skill – just as sow management, for example, is considered to be – time will tell.

In the meantime, thanks are repeated to Nick White, manager to Mr Roger Mercer in Staffordshire, for help in deriving the original text and to Prue Gillbanks for generating the layout for the book. As are acknowledgements to Pigtales Limited for the use of certain items of data, Graham Page for permission to take photographs and John Foster for actually taking them!

Figure 11 is based upon Figure 2-4 in Colin Whittemore's *Pig Production: the Scientific and Practical Principles*, and is reproduced by kind permission of Longman.

In revising the 1982 text my long-time friend Roy Holmes, Pig Director of Ro-Ro Farms Limited, Branston, gave me helpful opinions and my wife spent several hours in deciphering my handwriting so that the publisher might stand some hope of understanding it. My thanks to them both are recorded.

Newton-by-Toft   GERRY BRENT
Lincolnshire
1987

USING THIS BOOK

THERE will be those who will pick up this book and declare that it does not contain the solution to the particular problem with which they are grappling. Of course, this is quite possible as new circumstances arise in livestock production almost daily.

However, experience suggests that the most likely answer to pigmen's problems *will* be found in the book. This is because most difficulties have their origins in husbandry. If that is not true in every case, one other thing is sure to be. That is, no matter what is the cause of unsatisfactory performance, there will be a need for total commitment by the pigman to adapt, or review, his actions in order that the full benefit from any change in feed, medicines, stock or buildings is to be achieved.

As declared in the preface of this book, a pigman's lot can be a rather lonely one. However, it is important not to become dispirited and believe that only by the intervention of some non-husbandry measure can improvement be made. By the same token, it is equally necessary that the merit of an altered practice is not clouded by a failure to follow closely the agreed procedures.

The pigman does not only have an often lonely job, he has a very influential one; influential on the success and profitability of a unit, that is. It is acknowledged from the outset that it is not easy to evaluate the effect on performance produced by changes made. We all know of 'mystery' changes which appear to come and then leave of their own accord with apparently little input on behalf of the pigman. Fair enough, but *something* caused the changes and, if they persist, something *has* to happen to normalise matters.

It is at this point that the mental attitude of the pigman is vitally important. He has to ensure that he is not behaving like one of the two extreme types all too commonly seen.

Type 'A' hears or reads of a good idea and dashes off home to try it. He does not record when he made the change, or what else was adapted at the same time. Result?—no one really knows, and a useful amendment may be cast aside or lost by which time this pigman has rushed off on some other 'wheeze'.

Pigman Type 'B' is different. He will not use a new suggestion because it 'wouldn't work on his unit' or 'he tried something like it once'. He will stand back and complain about his lot and expect a new investment or a new 'wonder drug' to get him out of his troubles.

If we neither rush in nor close our minds to change, but instead review what is happening now and systematically recheck what we have done, how and when we do it, then improvements—very gradual at times—normally begin to happen. Just because something was tried a few years ago does not mean it is not worth another 'go' now—as stated previously, new circumstances continuously occur on a farm which has livestock.

So, when turning the pages of this book it may appear to the more experienced eye that it contains nothing new, and that may be so. However, if these words encourage the operator, whether he has a problem or simply wishes to achieve still greater efficiency, to stop and reconsider his actions then that pigman is halfway to what he wishes to achieve. If the reader can honestly say that he has worked his way through the various checklists in this book then it is likely that some improvement will arise. It will need only one change to prove the two main points of this opening chapter: firstly, that every single operator—new or experienced—will become more effective if he regularly reviews what he does and believes. Secondly, the pigman will be shown how great an influence he has on the unit's performance by the powerful effects produced by the changes he has initiated.

SECTION I

Chapters 1–6
Organising the unit

Including checklists for:

Typical daily routines in unit operation
    with practical examples given
Weekly routines
Routine checks upon equipment
Recording the unit
Emergency procedures
Moving pigs
Ventilation equipment
Pressure washing.

Chapter 1

LIKE ALL living creatures, pigs are never completely predictable in the way that they will perform or behave. Add to that the fact that each pigman is expected to tend an ever-increasing size of herd and it will immediately be seen why we need to establish a sound routine. Failure to organise oneself and those who are junior in the team can lead to less effective use of the human resource and a failure to observe early signs of that very unpredictability of pigs.

Pigs rapidly accustom themselves to a particular routine. This becomes an invaluable aid to the stockman who can simply measure the pig's response to that routine. The converse is also true in that if a major change in daily routine occurs – particularly in respect of timing of feeding – those animals which have to wait for an exceptional period of time may respond badly to broken routine.

Thus, a soundly planned sequence of work will not only ensure tasks are completed more easily and more quickly but it will lay the foundation by which operators can judge which animals are normal and those which require particular attention. A good routine helps both stock *and* stockman.

Examples of two basic routines are given. Whilst they could not be precisely followed on every unit due to structural differences, they allow the considerations which must be made to be illustrated in drawing up routines on any pit unit to ensure that key tasks are not omitted and that time and pigs are effectively managed.

DAILY SERVICE AND DRY SOW HOUSE ROUTINE

TASKS
First   Feed all occupants of house(s)

Note   *Where once-daily feeding is practised those animals requiring more than 2.5 kg of feed per day (i.e. newly weaned sows, some gilts prior to service and boars), should have additional feed given after service routine has been carried out in the afternoon. It is important that the same sequence of feeding is followed each time to avoid the effect of over-excitement in sows forced to wait longer than usual.*

5

Second

While animals are standing, clean out the sows, remove muck and bedding.

Repeat actions in boars' pens (may not be daily operation).

Third

Immediately after cleaning-out sweep access passageways and check sows and boars closely for:
- failure to eat ration,
- signs of unusual behaviour which may indicate oestrus or bad health,
- indications of lameness,
- vulva discharges which may be signs of oestrus, possible abortions or imminent farrowing,
- variations in sow condition.

Note   *The operator should mark the animal or pen where any actions which give rise for concern are seen so that the appropriate action may be taken later. Where urgent action is indicated the animal should either be removed or the advice and assistance of senior personnel sought.*

Fourth

Check house equipment—includes watering facilities and temperature control equipment if fitted, and also manure-removal equipment or slurry levels to help plan the rest of the day's work.

Fifth

There is much to commend that the more senior staff member undertakes service house operation due to its vital importance in achieving maximum sow output. If the person undertaking service routines is responsible for the decisions on boar:sow/gilt pairings and the number of gilts required to be served, the Fifth task could be the serving and checking of sows (see text for Sixth task—below).

Where the operator in the dry sow area does not make the decisions, the Fifth task will be to consult with the person with that responsibility so that matings can be planned. At the same time actions and decisions following the animals observed during earlier routines can be planned, treatments agreed, veterinarian called and so on. It is advisable to record any treatments given in the appropriate part of the unit recording system so that reference may be made about herd problems whenever relevant.

Sixth Check for oestrus and serving of sows and gilts.

Note *The only successful check for oestrus is to place gilt/sow*

and boar together, not mere inspection of the sow in her
pen. (For more details see Chapter 10 on service routine.)

Accurate recording of all services must be made (see
Chapter 4).

Any gilts observed in oestrus, but which are not to be
served on this occasion should be recorded so that they
may be fitted into the service programme at the desired
point.

Seventh Between the morning and afternoon service pro-

grammes the normal unit tasks may be conducted. These
tend to be a mixture of daily, weekly and occasional
tasks and will tend to vary, therefore, from day to day.
Circumstances may demand the movement of stock in
and/or out of the dry sow area. Muck removal, feed
mixing, mange dressing, cleaning down of pens, maintenance on building fabric or equipment, pregnancy
detection.

Eighth Repeat the Sixth task, checking and service routine and
recording.

Note *This should be carried out as late in the day as size of*

herd permits, in order to make it the last but one job in
the dry sow area.

Ninth Feed extra feed to those sows and boars on higher feed

levels. Check water, ventilation equipment, stock noted
under Third task of day. Fill feeders or barrows for
morning feed.

Check gates and doors are secure.

Night Night-time checks may reveal early signs of oestrus and
illness. Equipment should be checked again and the
house(s) secured for the night. This check during a quiet
time is very valuable.

DAILY ROUTINE IN THE GROWER/FINISHER HOUSES

TASKS

First Inspect all pens for trough/floor cleanliness and appearance of stock.

Note *If automatic feeding system is installed, check that empty pens are by-passed.*

Second Feed pigs—if not automatic, weigh or measure appropriate quantities and distribute evenly. Calibrate feeders.

Third Check all pigs, and mark or remove any not feeding normally.

Fourth Check water and ventilation equipment, adjusting as necessary. Check that feed distribution equipment cuts off correctly (if installed).

Fifth After discussion with whoever makes the unit decision, treatments may be undertaken and, if necessary, veterinary treatment sought, post-mortems considered and any such activities recorded in the unit recording system.

Sixth Remove muck, add bedding where applicable and sweep clean passageways.

Seventh Between the morning routine and the afternoon feed and observation session, the irregular routine tasks will be carried out. In the grower/finisher section these may include moving pigs, loading for sale or slaughter, weighing, pressure washing, muck and slurry disposal.

Eighth Repeat of Tasks one to five from morning routine as late in working day as convenience dictates.

Note *Where feed is offered more than twice daily, full effectiveness is achieved only when approximately six hours elapse between feeds. In this case, the third feed of the day should be given when other unit sections are given their Night Check.*

Chapter 2

THE NEED for a sound weekly routine becomes even more essential when more staff and more pigs are involved on a unit. Careful planning is also an aid to good herd health through the establishment of a sound hygiene programme. It is highly desirable to plan weekly events so that pens are clean and dry before reoccupation and so that they are used to full capacity without compromise on this important part of the routine.

There are possibly two key pieces of information which need to be known or decided upon before the routine is established. Firstly, as a key to the movement of pigs and cleaning routines, the day(s) on which pigs are despatched from the unit must be known.

To some extent the decision on the precise day of the week when weaning takes place will have an effect on the anticipated workload. For example, if sows are weaned on Tuesdays there will tend to be peak service/checking for service activity from the following Sunday onwards. Clearly, weaning on Thursdays would tend to have the effect of shifting the day on which checking of that group of sows would commence. In practice, however, checking sows for signs of oestrus is affected by the normal spread of time over which this occurs in the individual sow and, in any case, there will be less predictability as to the time when maiden gilts commence 'heat'. However, the pigman should be aware that he can influence, to some extent, the weekend mating workload by adjusting the day on which sows are weaned, and should take this into account in planning his weekly routine.

Thus, assuming a Tuesday collection of pigs for slaughter and a once-per-week weaning routine, on a breeding and feeding unit a typical weekly routine—to fit into that period between standard morning and afternoon routines outlined in Chapter 1—might be as follows:

Monday
Breeding herd Move sows served four weeks ago from service zone, wean litters (remove slurry), pressure-wash emptied farrowing pens.
Feeding herd Weigh and mark pigs prior to despatch.

Tuesday

Breeding herd Supervise farrowings, prepare pens and equipment in pens washed Monday.

Feeding herd Load pigs, relocate pigs either to allow growers to be moved into emptied pens or pens to be washed out.
Adjust feed scales of changed pens.

Wednesday

Breeding herd Supervise farrowings. Wash sows due to farrow next week and move to farrowing pen.

Feeding herd Move muck and slurry. (Move weaners into grower pens and pressure-wash weaner pens.)

Thursday

Breeding herd Routine piglet tasks—ear marking, castration, etc.
Supervise farrowings.
Administer prostaglandin for sows due to farrow on Saturday.

Feeding herd Move weaners into grower pens and pressure-clean weaner pens. (Move muck and slurry.)

Friday

Breeding herd Supervise farrowings.
Worm sows due to be moved to farrowing house next Wednesday. Pregnancy test sows.
Begin week's peak service routines.

Feeding herd Prepare weaner pens and equipment to accept weaners on Monday. Prepare feed and bedding stocks for the weekend.

Weekend

Breeding herd Peak of weekly service routine.

Feeding herd Routine minimised to morning and afternoon daily routines (see Chapter 1).

In addition to the tasks illustrated in this sample day-to-day weekly routine, there are weekly tasks which have to be considered by the pigman. These may be on varying days because they are not necessarily relevant to the typical sequence shown. They are, however, essential to the smooth operation of the unit and must not be overlooked.

- Check feed stocks and place orders.
- Check fuel stocks and place orders.
- Check contents of the medicine cupboard as well as disinfectants and detergents
- Check tool and hand implement maintenance as well as light bulbs and other equipment.
- Check records to assess the need for gilt transfer and selection.
- Check the readiness of isolation premises for incoming stocks.
- Check unit alarm systems and fail-safe devices.
- Check pigs ready for despatch the following week and advise outlets.

Chapter 3

PIGMEN CARE about pigs and for pigs. Pigmen do not care for machinery and equipment—well, perhaps, this one *is* a little sweeping. It is, however, very important to consider tasks which should be part of the pigman's routine, but which have only an indirect effect on the stock itself, perhaps at considerable intervals of time.

At this point it is proper to consider those tasks requiring regular attention but which are frequently forgotten or overlooked. It is essential for all those working with stock to consider what effects failure to undertake routine maintenance tasks will have. It is not just a matter of inconvenient breakdown or failure of equipment. Lack of attention to water and feeding fittings can—as indicated below—have a much greater impact upon pig performance.

So, by caring for equipment, we *will*, ultimately, be caring for our pigs. We will also be making life easier for ourselves in the long term and could be extending the effective life of the equipment which we use.

BULK FEED BINS AND AUTOMATIC FEEDING EQUIPMENT

During the 1970s, a rapid acceleration of the change to bulk delivery and storage of pig feed took place. It is easy to arrange for the delivery vehicle or system to top-up bulk bins but much more difficult to inspect them inside. It is so inconvenient to get them empty that it is often the case that they become a source of concern to us before we realise the implications.

Problems can arise from varying sources. Bin design may not be good. Those bins constructed without a 'breather' pipe are prone to problems arising from condensation droplets. Occasionally feed can be transferred into a bulk bin too hot from the pelleting/cubing process. These two circumstances stemming from 'physical' problems will have a similar effect on the contents of the bin. They cause the heating of pockets of feed or indeed the whole bin contents. When a mass of feed heats up it tends to draw moisture to itself and then, if and when it cools, the feed tends to stick together. These are the causes of 'lumps' of feed being seen in feed barrows or hoppers.

The problems will not stop there. Due to the heating of the feed, bacterial growth will be more rapid. In other words, those parts of the bulk bin which are turning feed 'mouldy' are also providing a medium in which organisms, potentially harmful to the pigs, can grow. So the problems arising from minor physical problems have now become chemical problems and the seriousness has multiplied many times.

It is also possible that some raw materials in the feed itself could be contaminated. There is growing awareness of a group of organisms, which can cause significant problems to pig performance, called mycotoxins. These will develop more rapidly under bad storage conditions, as described above. Their mode of action is to create a complex set of conditions under which toxins—or poisons—are produced. These can upset the metabolism—or normal body functions—of the pigs. Indirectly, they can interfere with the availability of vitamins and minerals which may then set up a 'chain' of eventual problems. It is likely that some of those inexplicable 'flare-ups' of health or performance problems on a unit can be traced back to contaminated feed and that, in itself, can be traced back to poor feed bin management.

The problem is that once a bin has become contaminated with mouldy feed or bacterially reduced materials, it will affect any other feed placed in that bin. So it is essential that feed bins, automatic feed dispensers and delivery dispensers are thoroughly cleaned at regular intervals.

The cleaning of bulk bins is frequently difficult and many bins have been constructed without inspection hatches. If older bins with no access points are used, access points should be inserted so that inspection and cleaning may take place. Although this can be achieved at low cost, the advice of the bin manufacturer should be sought to ensure that bin strength is unaffected. Specialist cleaning services exist in almost all parts of the country and it is usual to hire such a service to remove all material from the inside of the bulk bin itself and leave it in a clean condition. The pigman should ensure that all equipment within the houses themselves is emptied and clean.

The organisation of this work will entail:

● Allowing the bin to run empty to coincide with the arrival of the cleaning service.
● Cleaning the bulk bin and all equipment within the house.
● Allowing one week following cleaning for all equipment to dry before being re-used.

Despite the inconvenience of not having use of the bulk bin for a period and the possible need to handle bagged feed for around two weeks, this ought to be built into the routine on every unit. It may be sensible to spread the emptying of bins on a unit over a six-month period so that only one or two are emptied at a time which may cause less disruption to the unit operation.

In any case bulk bins should be emptied and cleaned at intervals no longer than six months.

MILK BY-PRODUCT TANKS AND EQUIPMENT

It has long been understood by users of skim and whey that there is a special need to clean regularly tanks which receive and store these materials.

Washing down the inside of tanks between deliveries is essential to minimise bacterial build-up. The best way to allow for this is to establish a two-tank system for both types of by-product. While one tank is washed clean the second 'overlaps' by being filled at that interval which guarantees that supplies do not run out.

The use of additives (usually formalin at the 1–1½ litres per 1000 litres level), will help to ensure that the product remains in a 'fresh' state with the minimum of adverse bacterial development, although some suppliers add preservatives on behalf of their customers.

WATER SYSTEMS

Another widely overlooked source of sub-standard performance is the drinking system. There are many hazards and it is possible that too many of these exist on a farm.

A pigman cannot be held responsible for a badly designed or faultily installed water supply layout. However, he must ensure that such a layout is not the cause of sub-standard pig performance.

This requires the pigman to check water supplies and this is particularly important in grower/finisher houses.

A common fault is that when there are peak demands for water at feeding times, the water pressure is too weak for the pigs to obtain sufficient. This depresses appetite which depresses performance.

So, as a part of every feeding routine the water supply should be checked. A very simple practice is to use a small container—a plastic cup will suffice—and note the time taken to fill the container at each end of the building—nearest to and furthest from the header tank. An adequate 'head' or rate of flow will be indicated if there is little time difference between the two points.

If there *is* a discrepancy, the routines for checking the water supply outlined in Chapter 16 should be followed. It might well be that the poor flow stems from gradual partial blockage of the supply lines to the drinker and not faulty installation. This is not uncommon and it is up to the stockman to prevent it.

Once it has been established that the supply, number and type of drinkers are all satisfactory, the pigman should:

● Check all drinkers on every feed occasion.
● Clean header tanks whenever the house or room is emptied.
● Ensure that header tanks are properly sealed.
● Ensure that water sterilents, if used, are incorporated to the manufacturer's instructions.

CLEANING AND WASHING EQUIPMENT

Much of the equipment used for muck and slurry removal and pressure cleaning has become more robust over the past decade.

It is essential, however, that the basic servicing requirements are followed. The problem with these important and regularly used items of equipment is that they are generally used by a number of operators and may not be attached to just one unit.

This means that it is easy to omit maintenance because no one person is responsible for that task.

A system known to work well is for one person to be responsible on the first Wednesday of the month (or whatever is appropriate to that particular site) for the maintenance of essential equipment which may include:

● Pressure washer
● Slurry tanker
● Below slat scraper
● Slurry separator and pump equipment
● Tractor.

VENTILATION AND HEATING EQUIPMENT

It is not necessary for the pigman to be a competent electrical engineer.

However, he must appreciate that equipment failure will arise if the contamination from the typical piggery is not reduced at regular intervals. Most of the items which need to be regularly considered

Plate 1. Regular checks on rate of water flow indicate whether this vital requirement is a limit to full performance.

Plate 2. Do not forget that the piggery equipment requires regular attention—like changing seals, washers and nozzles on pressure washers.

appear under the between-batch cleaning routine outlined in later chapters. In those houses where continuous stocking (e.g. dry sow house) is practised, the switching off, cleaning and lubrication of fans and calibration of control equipment must be carried out: **at intervals no greater than every six months, or as recommended by the manufacturer.**

Chapter 4

As WELL as his involvement in organising himself on a weekly and daily routine basis, the pigman also has a part to play in the strategic planning of unit operation. Overall decisions on the number of pens erected or number of sows kept may be made for, and not by, the pigman. However, once such policy decisions are taken, there is much that the pigman is capable of influencing which can have a large bearing on the effective operation of the unit.

Most units are planned to give a certain pig output capacity. From a business viewpoint, the decision to invest a given sum of money will have been based upon a predicted number of pigs passing through the unit. To plan such an investment, certain assumptions would have been made concerning the numbers of farrowings, litter sizes and growth rate of the growing and finishing pigs. Where these calculations are not met by results the proprietor is likely to find his investment placed in jeopardy.

At the practical level, failure to achieve the predicted throughput can bring a host of management problems to give the pigman extra difficulties. Simple, but important, examples of difficulties associated with a failure to organise and plan a well-regulated unit throughput are easily imagined and, unfortunately, quite common. In the breeding herd a failure to plan the gilt replacement programme adequately can lead to a trough in farrowings which will be followed by a subsequent peak. Consequently, at one moment there will be inadequate use of housing and equipment and at another so great a demand that boar usage, pen cleaning and weaning age might all have to be compromised in order to fit in the excess numbers of services and farrowings. In the finishing section failure to ensure that predicted growth rates are achieved will mean that pigs are either sold at lighter weights than budgeted, or that pens are overstocked which may lead to further husbandry problems.

Clearly these are matters which the operator can influence and most of the effective use of unit facilities rests with him.

In order that appropriate action is taken to ensure that pig throughput is maintained at the desired level, a comprehensive recording and monitoring system is required.

It is not uncommon to hear of a pigman being described as 'Good with pigs, but doesn't like keeping records'. Such a description is frequently due, in part, to a badly designed recording system which makes noting important details tedious and analysing the records almost impossible.

It does no harm to review, at regular intervals, the information being gathered and the manner in which it is gathered. There is no incentive for a pigman to maintain accuracy of recording where the details recorded are not used in order to make the unit more efficient. So design of a recording system which allows quick, simple notation on the unit and easy analysis in the unit office is a paramount need.

Not only will such a system permit more effective unit organisation to be achieved, but it will also provide the essential detail upon which decision-making can be based in order to achieve greater levels of efficiency. It is important to remember that no one can make judgements on the likely effects of changes in management practices unless there is a log of information on what is happening under the existing regime.

Nor is it enough to record simple chronological events like date of service, date of farrowing, date of sale. In order that the unit may progress, the performance of the individual animal and litter or pen of pigs must be monitored. In addition, details of any treatments administered must also be noted, so that the extent of any unusual circumstances can be assessed.

Having designed a data collection system which allows details to be simply gathered on the unit it is necessary then to adopt a method by which this information can be effectively used. The graphic representation of results has many advantages allowing at-a-glance understanding of any change or variation from prescribed levels of performance. Charts and graphs should be regularly updated and sited so that everyone involved can quickly see how performance levels are moving.

It is a good practice for the most senior operator to be responsible for updating such records, because it allows him to be in close and regular contact with the trends in each part of the unit.

There does come a point at which the responsibilities of the staff in looking after the animals threaten to become secondary to the upkeep of records. When that point is reached due to the size of the unit or to a high pigs-to-man ratio, outside agencies should be used to analyse the data and present results.

Up to the late 1970s most recording schemes run by the advisory agencies or commercial companies have been based upon the output and financial status of the unit. This provides a measure of the efficiency of a pig enterprise, but does little to guide the pigman, or his advisor, on why results are good, or bad!

So the usual unit measurements of pigs per sow per year, feed conversion ratio and cost per kilogram liveweight gain, help the owner to gauge unit efficiency, but not to correct faults or to become more efficient.

In each chapter on operational guidelines in this handbook step-by-step recommendations for the interpretation of results is given. At this point it is adequate to consider what information is required and effective ways of recording it.

Breeding Herd

To organise unit throughput, provide information upon which decisions can be taken and to monitor current output, the following details are essential on a continual basis:

- Dates on which sows are served
- Dates on which three- and six-week returns are due
- Actual conception data and/or abortions
- Dates on which sows are due to farrow
- The number of piglets born alive, dead and mummified
- Actual farrowing date
- Note of treatments given to sow and litter
- Fostering details
- Date on which weaning took place
- Number of piglets weaned
- Litter identification numbers
- Boar identification at service time
- Number of times sow served.

Effective recording schemes will also allow for occasional monitoring to be conducted in order to check performance over short periods of time. Included under these 'occasional' figures would be the following, together with any others required to investigate a problem, or monitor the effects of any changes:

- Birthweights
- Weaning weights
- Sow weights or condition scoring at farrowing and weaning
- Pens occupied.

If this data is collected on a form which contains certain performance guidelines, then the pigman may be able to make some immediate use of the records that he is asked to keep.

To achieve this it is desirable to adopt a system of target figures for the unit in question. This does *not* mean setting standards which are difficult to achieve. The best target figure is that which is needed to keep the unit facilities full to capacity, but not over-capacity.

The breeding herd figures should be divided into at least three sections and, preferably, four. These four sections are:

1. Service house details
2. Farrowing house details ⎫
3. Individual sow details ⎬ These may be combined.
4. Individual boar details ⎭

1. Service House Records

	Date of First Service	First or return	Sow or gilt No.	Empty days Date wea ned	Served by (boar) am pm am pm	Due to return 3 weeks 6 weeks	Pregnancy checks	Due to farrow	Remarks
1									
2									
3									
4									
5									
6									
7									
8									
9									
10									
11									
12									
13									
14									
15									
16									
17									

No. of services in previous 3 weeks ☐ Plus this week = 4-week total ☐

DIFFERENCE (+ or −) FROM 4-WEEK TARGET OF 56 SERVICES = ☐

NUMBER OF RETURNS IN PAST 4 WEEKS (TARGET 8 MAXIMUM) = ☐

N.B. Dotted line indicates weekly target number of services, reverse of card could carry individual boar service details (see Chapter 9). Sow age, or parity column might be incorporated.

Target number of services and returns set for individual unit. Some pigmen also prefer to provide space for the recording of sows weaned more than seven days, but not yet served.

The sample card allows adequate space for the gathering of essential details, plus instant recognition of any adverse trends.

From this service house sheet, the following can be calculated by whoever makes the decisions:

- Effective service/farrowing rate
- Weaning-to-service interval
- Boar work loads (see also Chapter 9)
- Future pen requirements.

2. Farrowing House Records

This sample card may be omitted by the pigman who may simply record the details on the individual sow record card (below). However, some collation of all weekly actions is needed so the

Farrowing house records

	Date farrowed	Sow or gilt no.	Served by boar(s)	Nos. born: A	D	M	Fostering In	Out	Mortality	Cull or keep	No. weaned	Remarks and treatments
1												
2												
3												
4												
5												
6												
7												
8												
9												
10												
11												
12												
13												
14												
15												

DUE TO BE WEANED IN WEEK = [　　　　]

NUMBER DUE TO RESERVE = [　] GILTS NEEDED TO ACHIEVE SERVICE TARGET = [　]

NUMBER OF PIGS WEANED = [　] VARIATION FROM TARGET (115/week) = [　]

N.B. Broken lines indicates target number of farrowings for one week. Code might be used to denote cause of mortality. Columns might be inserted for birth and weaning weights. Targets would be varied to suit individual herds. Age or parity of sow might be recorded to assist decision on culling.

stockman may plan his gilt service requirements to offset possible service shortfalls. A system known to work well, summarising essential farrowing house data is shown.

This farrowing house sheet will assist the decision maker to:

● plan future gilt service requirements
● decide upon sow culling
● monitor litter numbers, stillbirths and losses
● decide weaner pen requirements
● monitor treatments needed
● plot main causes of mortality.

3. Individual Sow Records
Ideally this card carries the lifetime detail of the gilt or sow and alerts the pigman to individual special characteristics of that animal. It is usual to use a double-sided card which contains details of previous performance in summary on one side and details of the current pregnancy/litter on the reverse. This card then is moved round the unit with the sow, permitting on-the-spot judgements to be made.

Historical sow record summary

	Sow's number									
	Litter no.	Date farrowed	No. born A | D | M			Fostered In | Out		Date weaned	No. weaned	Remarks
1 2 3 4 5 6 7 8										

N.B. If relevant, the parentage of the sow could be included along with the sow's date of birth and any test information. Details of sires of the litters and number of matings might also be shown.

This sow record allows the decision-maker to:

● assess the future of the sow
● predict possible problems such as difficult farrowing

- judge suitability as a foster-mother
- calculate average age of entire sow herd
- plan timing of culling to predict future gilt replacement needs.

The card which allows details of the current litter to be recorded on the reverse of the historical card is shown. It is usual to produce an updated historical card at weaning to go with the sow into the service house. Notice that this litter card example uses a 'box' approach for indication of actions taken, requiring a simple tick to be used when the task is complete.

Current litter card

Sow's Number						*Litter Number*		
		Boar				*Due*		
Date weaned	*Date served*	*am*	*pm*	*am*	*pm*	*3-weeks*	*6-weeks*	*Due to farrow*
RETURNED								

Born:

Wormed ☐   Farrowed   A ____ D ____ M ____

Teeth clipped ☐   Iron injected ☐   Ear numbers ☐

Fostered IN ____ OUT ____ DATE AND CAUSE OF LOSSES

Treatments and remarks:

No. weaned ____   Date weaned ____

N.B. Birth and weaning weight details may also be included, specific space for pregnancy checking and use of induced farrowing may be arranged.

This card allows the decision-maker to:

- check that tasks have been carried out
- assess the effects of certain treatments
- calculate the effectiveness of the sow against herd averages
- conduct an investigation into the reasons for piglet mortality

4. Individual Boar Details

The merit of monitoring the work rate and effectiveness of individual boars has become more apparent to pigmen in relatively recent times. The monitoring of a boar's working life is more difficult with the practice of cross-matings where one sow is served by two or three boars as an aid to increasing litter size. It is very desirable that each boar in the herd is allowed to repeat serve at least one sow every month to check fertility. To achieve this, arrange that the first service by each boar in any month is repeated on the same sow so that effectiveness of a boar can be measured. Where there is cause to question a boar's fertility, several repeat matings can be arranged quickly, so that a boar of lowered fertility can be identified and removed from the herd before causing widespread depression of sow output.

A two-tier method of recording boar usage is recommended. It is suggested that on the wall of each boar pen, the boar's identification

Boar recording card

Boar's number [ ]	Breed [ ]	Date purchased [ ]

Service details in week: Date of first service [ ]

	1 2 3 4 5 6 7 8 9 10 11 12 13 14 15 16 17 18 19 20 21 22 23 24 25 26
Year One	
Year Two	
Year Three	

Treatments and remarks:

N.B. Space may be included to record parentage of boar. Dates of swine erysipelas injections may also be included. If cross-mating is not practised, precise conception and litter size details from each boar can be monitored.

is clearly marked, together with a chart ruled out for each day of the week. From this is extracted—on the same day each week—the summary of number of matings carried out so that evenness of working between boars can be monitored. The individual boar recording card is shown on page 25.

These records allow whoever makes the decisions to:

● check the regularity of boar use
● decide upon culling programme and boar replacement needs.

FINISHING HERD

A recording method will provide information lacking in a large proportion of feeding herds. The benefit of recording finishing pigs is underlined in detail in Section VIII. In summary, the card, illustrated on page 27, will provide details which not only allow a series of management checks to be made but will also yield information which may permit the influence of changes in health, season of the year and the associated effects to be monitored. There has been a general reluctance to record finishing herds and this often means that there is a considerable deficit between potential and actual results on a unit. In a breeding/feeding enterprise it is common to find copious breeding herd records yet few for the post-weaning stages. The method shown will provide a basis for commencing finishing herd monitoring and provides opportunity to record:

● House and pen identification.
● Date, cause and approximate weight of pigs which die.
● Description of treatments given.
● Dates of check weighings and transfers.
● Weight and date of sales.

It may also be desirable to note feed scales, so that dietary changes can be assessed. The noting of slap marks for the allocation of grading details to varying management and housing methods adopted, can also be added where a rotary slapmarker is used.

From this, the decision-maker can:

● monitor the achievement of target weights
● plan selling policy
● predict feed requirements
● adjust management procedures

Finishing herd records

Pig age identification (week no.) [   ]   Sex [   ]

Grower house pen no. [   ]   Finishing house pen no. [   ]

	Date weaned	Check weigh	Date moved	Check weigh	Check weigh	Check weigh	Check weigh	Slapmarks and slaughter dates and weights
1								
2								
3								
4								
5								
6								
7								
8								
9								
10								
Age								
Target weight								
11								
12								
13								
14								
15								
16								
17								
18								
19								
20								
Deaths: (record date, weight and cause)								

Treatment and remarks:

● monitor effects of changes made on growth performance and gradings.

Summary

The above details contribute data to ensure the calculation of overall physical herd performance and, with a regular stock check, permit financial assessments to be made.

In addition, they allow the pigman, with guidance or the use of sophisticated computerised analysis, to spot quickly changes in trends and they provide him and his adviser with 'ammunition' on which may be based the vital decisions to alter the unit management practices.

Subsequent chapters deal with the vital consideration of record interpretation which relies upon the pigman maintaining accurate records out on the unit.

It is unacceptable for any pigman to be considered a poor record keeper. To be so is to weaken the effectiveness of whoever makes the decisions as well as his adviser in the quest for corrective and improved actions.

The more sophisticated, often computer-based, recording systems are no easier than any other recording method for the on-farm collection of data. However, they do make calculations easier, thus removing the chore of working out the effects of changes in performance and monitoring progress or identifying the need for any change. These record systems, by making data analysis easier, should allow the stockman more time to study them and to identify the strengths and weaknesses of his management.

Chapter 5

Throughout this book the aim has been to make everyone more aware of people and things around them on the unit and to become more efficient.

There are times, however, when on the best-run units things do go wrong. Some of them are outside the control of the pigman—and even his employer.

Every unit should at regular intervals review its emergency procedures. Every time a new employee joins the team it is to be recommended that the following routine is gone through. This not only allows the newcomer to be familiarised with the unit routine, but ensures that emergency procedures are reviewed at the same time.

Every pigman should know what to do in the event of the following typical list of emergencies.

1. Power Failure

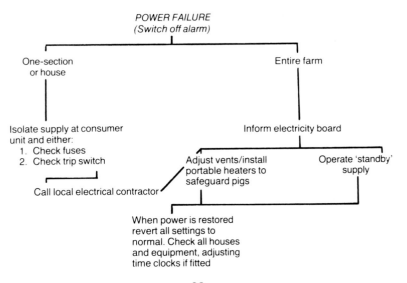

29

In order that anyone can follow such a routine, they need to know:

1. Position of all fuse boxes/consumer units.
2. Procedure for replacing fuse.
3. Number of local electricity board (should be prominent, near to phone), and local electrical contractor.
4. How to operate emergency standby unit.
5. Location of emergency heater units.
6. Procedure for checking security of stock and equipment in all weather conditions.

2. Water Failure, or Burst

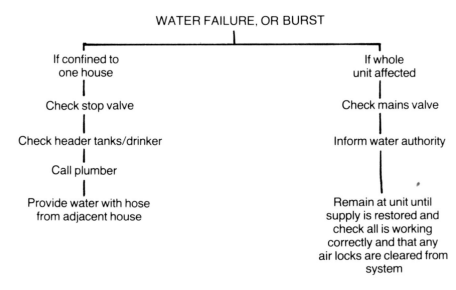

WATER FAILURE, OR BURST

If confined to one house — Check stop valve — Check header tanks/drinker — Call plumber — Provide water with hose from adjacent house

If whole unit affected — Check mains valve — Inform water authority — Remain at unit until supply is restored and check all is working correctly and that any air locks are cleared from system

In order that anyone can follow such a routine, they need to know:

1. Location of mains and stop valves.
2. Procedure for checking drinkers.
3. Number of water authority and plumber (should be prominent near to phone).
4. Location of troughs and hose.

3. Fire at Unit

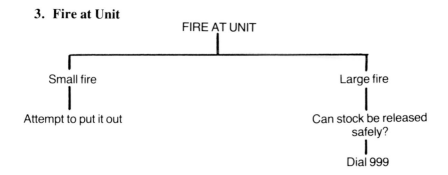

In order that anyone can follow such a routine, they need to know:

1. Position of fire fighting equipment.
2. Suitability of various extinguishers for electrical fires (dry powder extinguishers) and structural fires (normally water).
3. Emergency exits for stock.
4. Fire hydrant points for fire brigade.
5. How to give clear location directions to fire brigade.

4. Adult Boars Fighting

In order that anyone can follow such a routine, they need to know:

1. The whereabouts of would-be assistance even at weekends or at night.
2. That they should not attempt to separate the boars single-handed.
3. The location of boards or hurdles.
4. That excited boars may act aggressively toward staff and be quite 'out of character' as a result of a fight with another boar.

5. Animals Trapped Under Gates, Crates and Divisions
In order that such pigs can be released everyone needs to know:
1. The availability of assistance even at weekends, or at night.
2. Location of crowbars and tools.

6. Accident
In order that prompt attention can be given, everyone needs to know:
1. Designated unit first-aider.
2. Location of first-aid cupboard.
3. Telephone number of doctor and hospital.

PIGMAN'S EQUIPMENT

Although not always required for emergencies, a checklist of equipment is now given, so that a pigman is prepared to undertake unexpected, or routine stock tasks whenever they require attention. This means that *the pigman's cupboard* should not only be stocked, but regularly checked for supplies of all essential replaceable items.

Scissors	Marking sticks or sprays
Gut (sutures)	Ear notchers
Surgical needle	Tattooing pliers, numbers and
Tail docking forceps	ink
Tweezers	Restraining rope (snare)
Scalpel and blades	Muzzle for savage sows
Syringes and various sizes of	Teeth clippers
needle	Anaemia prevention equipment
Thermometer	Tranquilliser
Antiseptic solution and/or	Medicines supplied by
powder	veterinarian
	Cotton wool

The following should be available for emergency treatment of stock:

● Trays and troughs for feeding small or orphaned pigs.
● Glucose, milk substitute and electrolyte additives.

Chapter 6

It is when some of the more arduous tasks on a pig unit are being undertaken that it is possible to detect the more experienced operator. It is one who thinks ahead a little to plan the job and makes sure that difficulties that have hampered him before do not crop up again.

It is not good enough for the pigman to complain that the layout of the farm makes jobs difficult. Buildings cannot easily be moved and there may be sound reasons for buildings being placed in a particular position which, although logical in their own right, make practical operation less than perfect.

To cope with such circumstances it is necessary for the pigman to devise an approach which can be used to overcome such problems—as underlined in Chapter 22.

In this chapter three common, almost everyday unit tasks are considered and the 'thinking' pigman's approach to the tasks is suggested.

Movement of Pigs

The primary consideration must be that the pens to which the pigs are being moved are in a suitable condition to receive them. This underlines the importance of a sound weekly routine and the use of a recording system which permits advanced planning to be made (see Chapter 2).

Now for the task itself

Prepare any equipment required, such as moving boards (no sticks!). (Movement boards are ideally constructed from 12mm-thick plywood with handles cut 75mm from the top of the board which should be 1.2m high and 100mm narrower than the unit passageways.) Weighing machine calibration should be checked and recording cards and pencil prepared as appropriate.

 Open gates and doors of pen and house into which the pigs are to be moved. Ensure drinker and feeders are working and that bedding, if appropriate, is available.

Remove any obstacles and tools from the route which the animals are due to take.

Erect barriers or protection gates so that 'escape routes' from the course that the pigs are due to take are closed.

Remove any equipment from the receiving house and the house from which the pigs are being moved which might be damaged, such as hoppers and heaters.

Alert helpers so that sufficient assistance is on hand to make the task as easy as possible.

Remove pigs from their pen; close pen, room, house and passageway gates at each appropriate point to prevent the pigs rushing back to their original point. Boards should be used which allow knees to be placed against the board to present a solid obstacle to the reluctant animal. Never get in front of the pig unless you wish to make it turn or stop, and watch the pigs closely for signs of reluctance.

Record the movement, ensure feeding and management instructions are clear to everyone who may be responsible for those pigs.

Day-to-day Checking of Fan Ventilation Equipment

In any house where fan ventilation is used, the temperature and the equipment must be checked at least once a day.

It is vital to establish one point which is commonly misunderstood. Most ventilation control panels have a set temperature dial. The object of this dial is to allow the operator to set the temperature which is desirable for the pigs in the building. It must *NEVER* be assumed that the temperature set on the dial is being achieved in the house. It is strongly recommended that the stockman assumes that that temperature in never achieved. So some separate checking should be made. Furthermore, the temperature sensors within the house, which pass the message back to the control panel, will be placed in an arbitrary position and will never precisely represent the temperature at pig level.

So the pigman should start with the basic assumption that it is his interpretation of the environment within a building, *not* the provision of automatic equipment which will make the difference to the comfort and, ultimately, performance of the stock.

Now for the task itself

Inspect the pigs—preferably between feeds. Do they lie in a comfortable manner? They should be happy to recline on their sides, stretched out and touching each other in the area of the pen designated for lying. If they are scattered and fouling the lying area, they may be too warm. If huddled together and 'complaining' by squeals and moans, they may be indicating that temperature is too low.

Check the temperature on the recorder. This might be:

1. A thermograph recorder which senses the temperature and inks the measurement on a drum revolving on a clock mechanism. While these are ideal, they are costly to provide for each house.

2. Maximum/minimum thermometer suspended adjacent to the temperature sensor or positioned in several places within the house. In a new house, or any house giving problems, the readings should be noted on a prepared recording card or book and the control panel setting recorded alongside.

3. A hand-held digital thermometer which is moved and allows 'spot' temperature around the house to be monitored and, preferably, recorded.

The sensor, which will normally be either a thermistor (for variable speed fans) or thermostats (for on/off 'stepped' fan systems) should be checked. It is important that these sensors are maintained in a dust-free status, but they are not designed to withstand heavy-handed cleaning. A gentle jet of air is the best means of cleaning.

Check the settings on the control panel. Ideally, adjustments should be made only in conjunction with the unit decision-maker and any adjustment should be noted so that the effect can be properly measured and experience gained. It is unsatisfactory and very con-

fusing if ad hoc adjustments to control panels are made
by a number of staff on the unit.

Return to the house and check temperature and pigs
again four hours after any adjustments to the settings
have been made because it will normally take two to
three hours for the ventilation system to have some
effect on its revised setting.

For fault-finding routines on temperature control, see Chapter 7.

PRESSURE WASHING

The need for good unit hygiene is discussed in detail in Chapter 8. It
is not generally realised by pigmen that at least half of the time
taken to effect the pressure washing of a pen or house is involved in
setting up for, or cleaning up after, pressure washing, not the act of
cleaning itself. This makes a routine even more important.

The advent of very high pressure washers in recent years has
reduced, considerably, the time required to lift and remove muck
from the surfaces of a pen or building. However, these intense jets
of water must be directed with care because they increase the
likelihood of damage to electrical fittings and the fabric of the
building including insulation materials and even concrete floors.

Now for the task itself

Remove all portable equipment from the house, such as
heaters, feeders, pen divisions and recording cards.
Enclose electrical fittings in polythene or mask them
with sealing tape, and isolate mains supply if possible.

Carefully scrape and brush any loose muck, feed or
bedding from the area to be pressure-washed and
remove from pen and house. Ensure that house drains
are clear, but that drain grids are in position.

Set up washer, ensuring that electric cable is
positioned securely and fixed to comply with Health and
Safety Regulations where there are no collections of
water. The filter on the water suction pipe should be
checked to ensure that there is no blockage which might
restrict flow and no kink in the pipe, also that the filter has
no holes that might cause the nozzle to block or the pump
to be damaged. Ensure that the water supply is adequate
but avoid wasting water by over-filling the washer sump,

because this adds unnecessarily to the unit waste system.

Soak all pen surfaces—preferably using a detergent.

Soak all fittings. Ideally this is achieved outside the house in specially constructed 'dip'.

Return and wash all pen and house surfaces. Start with thorough cleaning of any fixed fittings, such as farrowing crate or trough, then clean the floor, then the walls and divisions, finishing with a comprehensive rinse of all surfaces.

Then wash all tools, equipment and removable fittings.

Disinfect all surfaces, if disinfection is part of normal routine. Many pressure washers are not designed for the use of coal tar and phenolic compounds, so a separate knapsack sprayer may be safer to use for this task. Pressure-washer jets may become blocked and it is difficult to calibrate them precisely to achieve correct dilution rates for disinfectant unless special attachments are provided. Always make sure that there is no residue of disinfectant around before rehousing pigs.

Stow pressure washer and equipment, commencing with electrical flex and water hose. Ensure that the washer is stored so that it is protected from frost during the winter months.

If fumigation is practised, it may be carried out at this point, paying great care to essential safety considerations. Place fittings inside the house or pen and allow the house to dry thoroughly before installing equipment in ready-to-use position and checking that all is serviceable. All ventilation equipment should be carefully checked at this stage, including back draught flaps and inlet vents.

Although these are just three of the regular daily tasks, the principles involved in conducting the work may be applied to a whole range of work. The principles are:

1. Prepare the job.
2. Move smaller items to prevent damage.
3. Do not ignore safety points.
4. Leave everything ready for the next job or use.
5. Record actions taken so you can check effectiveness of what you have altered.

SECTION II

Chapter 7

Providing a better environment

Including checklists for:
Influences upon animal comfort
Correcting adverse temperature
Effects of incorrect house temperature
Requirements of various classes of pigs
Air movement within a building
Using air movement principles
Checking the system.

Chapter 7

THERE CAN be no one in our industry who does not understand that the correct temperature for the varying classes of pigs is of vital importance. Yet this remains one of the poorest-managed components essential for satisfactory pig performance.

It is usual to think of inaccurate temperature control as a major contributor to poor feed usage, bad dunging habits and poor pig growth.

It is inaccurate to limit considerations of temperature control to these factors. Temperature control can have far wider-reaching effects upon herd health, feed uptake or appetite, reproductive activity, survival rate in new-born piglets and a whole host of factors which will be discussed under the problems outlined in later chapters.

There are many components which may influence house temperature control but it is fair to say that the one input which can exaggerate or mitigate the effects of all the remaining components is—the pigman.

The standards for ideal temperature requirements of the pigs can be found in most pig textbooks. It is important to remember that the pig's response to temperature will vary with other conditions to which it is subjected. The example which follows indicates how several interrelating factors can influence the effect of house temperature upon the animal. It also serves as a framework for the pigman to consider those actions which he might take to minimise the effects of adverse conditions upon temperature.

An In-Pig Sow:
housed in a stall
on a partially slatted floor surface
in a large building with fan ventilation and hand-adjusted air inlets
with some draughts
being fed 2 kg per day of a medium-density diet
will perform below her optimum if the house temperature is below 18°C.

BUT If In-Pig Sows are:
penned in groups
in a pen where deep-straw can be provided

where the lying zone is covered to reduce air space
where draughts are eliminated by flaps and boards
where 2.4 kg per day of a high-energy diet is fed
they will perform below their optimum if the house temperature is
below 9°C.

Lessons for the stockman from this example

1. Interpret information with care. Do not bother to remember a
particular figure, e.g. 9°C in a sow's lying area, if you do not or
cannot *also* remember that that applies to generously fed sows lying
together in snug kennels with plenty of straw.

2. The pigman cannot alter the system—if stalls are provided,
group housing cannot be considered. Do not condemn individual
housing because it demands a higher house temperature. Stalls and
tethers have other advantages which normally include lower labour
needs and better litter size and no fighting.

3. You may not be able to reduce the airspace or add bedding,
but you can:

ensure that water spillage is minimised because water increases
the speed of heat loss from the sow and from passageways if their
surfaces are wet.

4. Ventilation control equipment can be recalibrated to ensure
that velocity is minimised to suit the conditions (see checklist
below).

5. Every effort can be made to draught-seal a building. No one
who may read this book would consider doing so sitting in a chair in
a draughty position—or at least they would be most uncomfortable
doing so. It is very difficult to ensure a perfect seal around every
door, window, vent and slurry channel—however, it should be
attempted!

There is a very close relationship between speed of air movement
and house temperature. Whether by intention (the ventilation
system), or by accident (draughts), it is likely that 60 per cent of the
temperature loss from a building is via air exchange, *not* through the
structure of the building. The temperature around the door or
window is usually 4°C below that in the main body of the building.

6. There is no sense in adopting a feeding regime which fails to
take into account the conditions provided for the pigs. The correct
combination of diet quality and quantity must be adopted to suit the
individual farm circumstances. Sow feeding should not be measured
in quantity alone, but in the pig output from that quantity or in
terms of cash response to the feed cost (margin over feed cost—see
Chapter 21).

Plate 3. All doors, windows, slurry channels and adjustable flaps should be sealed to prevent unwanted draughts occurring.

Thus, using the simple example on in-pig sows housed differently, it can be seen that:

● If we change the system, we change the 'rules' and we must interpret recommended standards for temperature requirements and feeding accordingly.
● Even though a pigman cannot alter a building, or improve its insulation properties, in many cases he can improve temperature control by attention to other details—draught exclusion in particular.

WHAT HAPPENS WHEN THE PIGS ARE TOO COLD?

Some of the critical effects of operating piggeries at too low a temperature have been mentioned in passing. It is now time to quantify the temperature needs of the pig and spell out in more detail the effects of falling below these levels.

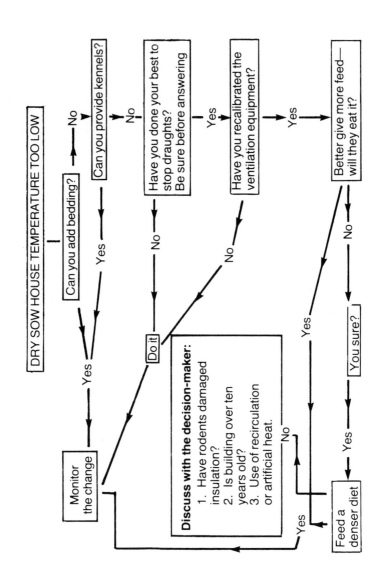

DRY SOW HOUSE TEMPERATURE TOO LOW

Can you add bedding? — No → Can you provide kennels? — No → Have you done your best to stop draughts? Be sure before answering — Yes → Have you recalibrated the ventilation equipment? — Yes → Better give more feed—will they eat it?

Can you add bedding? — Yes

Can you provide kennels? — Yes

Have you done your best to stop draughts? — No → Do it

Have you recalibrated the ventilation equipment? — No

Better give more feed—will they eat it? — Yes

Better give more feed—will they eat it? — No → You sure?

You sure? — Yes → Feed a denser diet

Monitor the change

Discuss with the decision-maker:
1. Have rodents damaged insulation?
2. Is building over ten years old?
3. Use of recirculation or artificial heat.

No / Yes

1. The Adult Sow and Boar
These need 12°C (groups and bedded) to 18°C (individual, unbedded and draughty).

What happens if it is below these levels?
- The body condition of the sows and boar may be difficult to correct.
- Poorer sow output may occur due to reduced sperm and ova fertility.
- Piglet birth size may suffer and so will survival rate.

What can the pigman do about it?
See page 44.

2. The New-born and Suckling Piglet
These need 30°C–26°C.

What happens if it is below these levels?
- Pigs huddle and grow less well.
- Weaker pigs are more likely to die due to depletion of body energy reserves.
- Pigs tend to lie closer to sow and chances of overlying are increased.
- Some pigs will not even have the strength to move to heater at birth.
- Failure to dry off rapidly after birth increases heat loss and chances of chilling.
- Lower temperatures are a major trigger of piglet scours and even respiratory conditions.

What can the pigman do about it?
See page 46.

3. The Newly-Weaned Piglet
These need: 26°C–22°C depending upon age at weaning.

What happens if it is below these levels?
- Pigs grow less efficiently.
- There is an increased likelihood of scour.
- Increased need to use drugs.

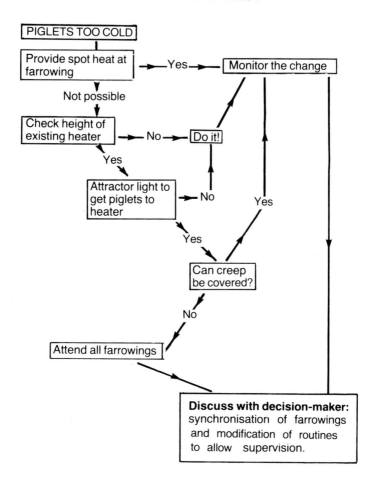

- Because of the increased risk of scour, there is an extra need to restrict feed intake which exaggerates the inefficient growth rate.
- There is an increased tendency to reduce ventilation rates which then increases moisture content, reduces extraction rate of airborne organisms and, therefore, increases the risk of respiratory problems.

What can the pigman do about it?
See page 47.

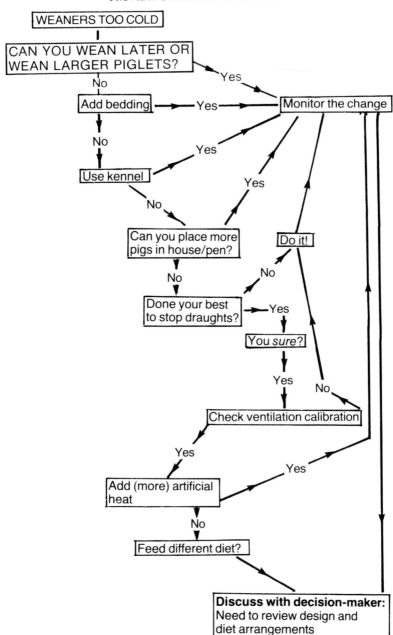

4. Growing and Finishing Pigs

These need: 22°C (at around 8 weeks of age in unbedded houses) to 18°C (90 kg pigs in unbedded houses).

What happens if it is below this level?

- Pigs in the range 20 kg–90 kg will need 25 g of feed per day on average for every 1°C below the desired temperature relevant to the conditions. Typically this means an increase, that is, a worsening of the feed conversion ratio of 0.1 which could be worth around £1.40 per pig depending upon diet cost.
- Put another way, if extra feed is *not* given, the pigs will grow around 11 g per day slower for every 1°C below optimum levels. Typically this will increase the time taken to grow from 20 kg to 90 kg by about eleven days.
- Both points (a) and (b) increase the feed conversion ratio by 0.2 due to the increased number of days the pig spends on the farm which increases the proportion of feed it needs for 'maintenance' rather than growth.
- All pigs react to a low temperature by increasing their bodily heat production which increases still further their feed conversion ratio.
- Pigs adopt unsocial lying and dunging habits if draughts are the cause of low temperature.
- If pigs are kept in colder conditions and fed moderately they will grow slowly and have low backfats; that is, they may 'grade' well but grow uneconomically.
- If kept in cold conditions but fed generously they may respond by a disproportionate development of fat, due to suppression of their lean tissue growth, to combat the low temperature and this will tend to increase backfat levels causing poorer gradings.

What can the pigman do about it?
See page 49.

WHAT HAPPENS WHEN PIGS ARE TOO WARM?

Although this is less of a problem in the United Kingdom industry, there are circumstances under which the effects of too high house temperatures are readily observed.

The pigman's reaction to these circumstances will be relatively standard, but the effects on the various classes of pigs will differ. Thus the likely consequences of too high temperatures are dis-

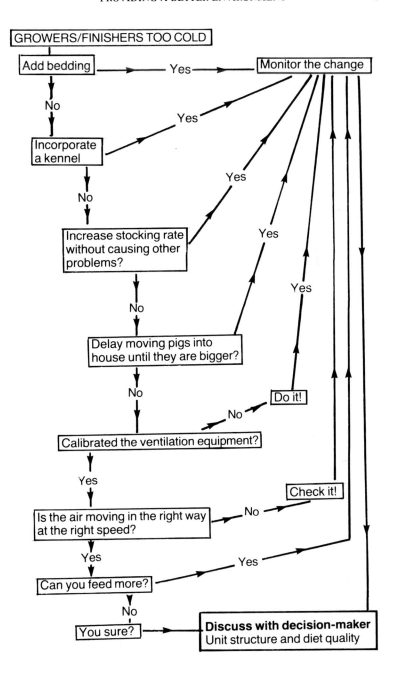

cussed under the various categories and one checklist is suggested
for the whole:

1. The Adult Sow and Boar
These need: Temperatures no higher than 18°C.

What happens if it is above this level?
- Sows become reluctant to eat high levels. This can be parti-
  cularly serious in the farrowing house because it can lead to sows
  being weaned in poor condition and milking less well.
- It can lead to heat fatigue which is particularly serious at and
  around farrowing time.
- In extended periods of high temperature, onset of oestrus after
  weaning is extended.
- Conception rate may decline due to depressed reproductive
  activity in the sow and low sperm counts in the boar.
- Boars become reluctant to work.
- If exposed to direct sunlight, sows and boars of white breeds may
  become scalded and this severely restricts their activities.

2. Suckling Pigs
These need: Temperatures no higher than 30°C.

What happens if it is above this level?
- Due to the fact that the piglets leave the sow at a temperature of
  over 35°C, need rapid drying and have few body reserves to
  combat cold, they are the category of pig most tolerant to high
  temperatures.
- One practical problem which may arise is that the piglets may
  adopt a very scattered lying pattern which may increase their
  proximity to the sow and raise the likelihood of overlying.
- If the creep area is super heated relative to the size of the piglets,
  they may choose to lie outside this area even if the rest of the
  piggery is below their optimum temperature and thus be affected
  by problems discussed under the effect of too low temperatures.

3. Weaner Pigs
These need: Temperatures up to 26°C.

What happens if it is above this level?

- Once again, this group is tolerant of high temperatures.
- Appetite depression can be a major problem, due to a slowing of

the animal's energy-absorbing activities and to the increased risk of feed contamination which may occur when in a trough or hopper at high temperature levels for extended periods.

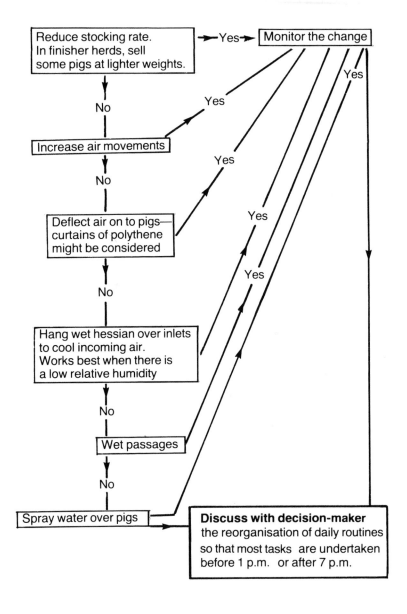

4. Grower and Finisher Pigs

These need: Temperatures of up to 22°C.

What happens if it is above this level?

- Pigs suffer acute appetite depression and, therefore, grow less well.
- They tend to utilise more of the feed to deposit fat; that is, they 'grade' less well.
- There is an increased tendency towards vices such as tail-biting.
- Pigs tend to 'wallow' in an attempt to cool themselves and they can 'manufacture' such an arrangement by fouling a solid floor section of the pen—this is often the 'lying' zone.

What can the pigman do about it?

See chart on page 51.

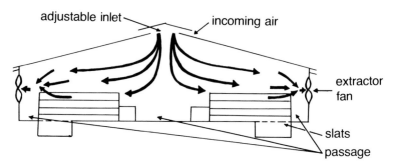

Fig. 1. *Assumed movement of incoming air*

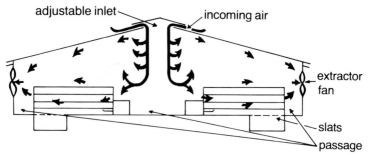

Fig. 2. *Likely movement of incoming air*

The Principles of Air Movement

There are several basic misconceptions about the manner in which air moves. While it is not necessary for a pigman to understand engineering principles, he should understand what happens to air when it enters a house, so that he can ponder the facts and consider how he can harness these patterns to good effect on his own unit.

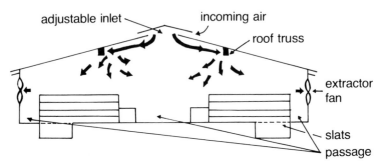

Fig. 3. Air currents disrupted by roof trusses

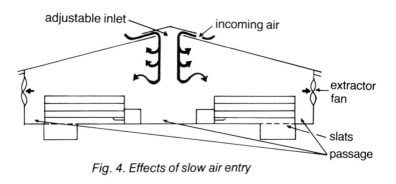

Fig. 4. Effects of slow air entry

Older textbooks used to illustrate air movement patterns like those shown in fig. 1. There might just be some circumstances where this is correct, but there will not be many days in the year where it applies.

Air will follow those patterns shown in fig. 2 *when* air is entering the house quite rapidly; that is, when winds are blowing or fans running fast, but it will *not* always follow this pattern.

In fig. 3, roof trusses are positioned so that they disrupt and deflect the air currents.

In fig. 4, the air is entering slower, so it does not cling to the underside of the ceiling, but falls like a 'waterfall'. This is a common condition where there is low ambient conditions and fans are running slowly.

How Can the Pigman Use these Principles?
These facts can be put to use to improve temperature control within a house. The points apply particularly, but not exclusively, to fan-ventilated structures.

The following points may be considered:

1. It is the manner in which air *enters* the building which dictates air movement patterns, not how it leaves the building.

2. The speed at which air enters has the greatest effect on how it moves around the building. Thus it is possible to 'throw' incoming air by 'choking' the size of inlet to match the fan speed.

3. Deflectors can be hung to cause incoming air to fall on to pigs.

4. Similarly, deflectors may be used to prevent air falling on pigs in cold weather or the size of inlets can be reduced to create the jet or 'throw' effect described in point 2 (above).

5. Because in most circumstances air is cooler as it enters a building, it is safe to assume that it is warmest near to the outlet; therefore, those pigs which respond to the highest temperature—the smallest pigs—should be placed closest to the extract or exhaust points if possible.

6. In low ambient temperature conditions it may be desirable to reduce ventilation rates to that point just above which condensation might occur. If this is achieved by closing off fans, then they must be isolated from the electricity supply to avoid inadvertent switching and damage, and the fan aperture should be shut off using a draught-proof slide.

Checking Ventilation Control Equipment
There are three main types of controllers for fan ventilation.
Manual controllers
These are a simple switch type which allow the pigman to operate a fan or number of fans as he decides the need arises. A variant is where he has choice of one to five speed settings.

The manual system relies heavily on the pigman's ability to measure the animal's needs and to adjust outlet and inlets according to the chosen ventilation rate.

Thermostat controllers
These allow a varying degree of automatic control. The pigman presets the desired house temperature and when that is reached the fan, or fans, run at their maximum speed to return it to the desired level. It is an automated on/off switch although it can be made more sophisticated by the use of a timer which allows the pigman to allow the set temperature to be 'over-ridden' for a chosen period at given intervals. This may be set at that level required to prevent condensation forming. Care with setting of inlet and outlet vents is required.

A further refinement is the use of a second thermostat which is wired to switch on supplementary heaters when the temperature drops below a preset level (normally 1°C below the desired house temperature).

Thermistor controllers
These are normally utilised for variable fan speed systems. The pigman is usually expected to set at least two dials. The first is set to the desired house temperature—as the temperature of the house rises above that setting, the fan speed increases from its minimum speed setting which is set by the second dial. Fan speeds should increase very rapidly when the house temperature exceeds the preset level to minimise house temperature variance.

Similarly, the fan speed should drop to the minimum setting within 0.5°C below the set temperature. Take care to set the control panel so that fans will reduce to 10 per cent of their fan speed. This still makes it difficult to maintain house temperatures at the likely extremes of their requirements, particularly if no automatic control of inlets or outlets exists.

In weaner houses and some farrowing systems, a further refinement is the addition of an interlocked heater arrangement to avoid temperatures falling still lower beyond the set temperature even with the fans at minimum speed.

Checking Manual Systems
The lying posture of the pigs will reveal the need to check house temperature accurately.

If pigs are *too cold* then the frequency at which the fan is running should be reduced (or fans switched off and sealed if more than one fan is installed). Care to adjust air inlets/outlets and seal doors should be taken.

It is possible to consider the use of reversing switches on extraction fans in cold weather. Although fans running in reverse may be

Plate 4. Before pigs are moved into a fan-ventilated house the pigman should check the setting and calibration of the control panel.

less efficient, this pressurisation may help low-temperature control.

If pigs are *too hot* the frequency and speed of fan operation should be considered and the fan checked to ensure that the diaphragm plate is sound and the fan blades not choked with dust. Pressurised systems do not exhaust effectively if doors are left open because air escapes without 'scavaging' foul air effectively.

Checking Thermostatically-Controlled Systems

Once again, the checks outlined in Chapter 6 may reveal the need for maintenance.

If the pigs are *too cold* the thermostat setting and accuracy must be checked. It may be necessary to vary thermostat settings to compensate for their arbitrary positioning within the house.

If the pigs are *too hot*, the similar set of checks as outlined for the manual systems should be followed.

Checking Thermistor-Controlled Systems

If temperature checks reveal a need for adjustments, the following routines may be considered.

If the pigs are *too cold*, after checking the settings, fan efficiency, inlet or outlet settings, consideration might be given to recalibrating the minimum speed setting which may be causing fans to run too fast. A simple pigman's check, preferably in a vacant house, is:

1. Turn set temperature dial to highest possible mark.
2. Set minimum speed dial to minimum and gradually increase until it is just possible to hear the fans running at a steady, constant 'hum'.
3. Then return set temperature setting to required level and check three to four hours later.

If the pigs are *too hot*, in addition to the checks underlined for other control systems a check should be made to ensure that the fans do actually reach their maximum speed and 'step' up quickly to full speed once the 'set' temperature has been reached.

Remember
Even where house design does not provide high insulation levels or accurate ventilation control, the pigman can still make a large contribution to house environment. A useful addition to equipment on any unit is simple temperature-monitoring and, preferably, air-speed-measuring implements.

It is unwise to assume that just because a dial on a control panel is set at a particular temperature, that temperature is being achieved – that is where monitoring equipment can give greater accuracy and can help to illustrate where draughts occur. Such equipment also allows more accurate assessment of the environment at pig level which is typically quite different to that at the point where the temperature sensor is positioned.

SECTION III

Chapter 8

Pig health and unit routine

Including checklists for:
Disease prevention
Use of the veterinary surgeon
Keeping disease out
Problems associated with incoming pigs
The acclimatisation routine
Problems associated with vehicles
Reducing problems associated with visitors
Reducing problems associated with birds, rodents and other
  animals
Hygiene as a routine
Specific conditions: mange, erysipelas, worms.

Chapter 8

AN ONLOOKER at many herds might gain the impression that inside any pigman is a veterinarian waiting to get out! The converse may also apply—frustrated by what they see in practice, many vets find that the needs for treatment arise because of defects in stockmanship, so they spend much of their time thinking and talking about pigmanship! The veterinary profession has been preaching the advantages of preventive medicine for many years now. Unfortunately, many pigmen have not heard—or heeded—the message. They still look upon illness as something for the vet to treat, *not* for them to prevent.

Disease outbreaks occur through two major channels. Firstly, when a large dose of an organism is 'imported' on to the farm. Secondly, where the organism is already on the farm, but some change in practice allows the level to increase and illness to occur. Clearly, the pigman has a part to play in preventing either circumstance.

Gradually the impact of health as a component of success or failure of a pig business is being realised and many herds now use the veterinarian as an aid to management, rather than an agency attempting the cure of disease. It is not enough simply to invite the vet to 'walk the farm' every so often. If the vet, and the pigman, are to understand the trends within a herd, a sound recording system must be maintained and the results made available to the veterinarian and adviser. Only in this way can the impact of any husbandry effect on unit health be measured.

The records, outlined in Chapter 4, allow for treatment, illness or poor performance to be recorded. Prior to a consultation with the vet or adviser, the key performance factors should be analysed so that they might be compared with the output in previous periods. Coupled with the herd inspection, this will permit a much more effective use of the adviser's time.

ORGANISING THE ADVISORY VISIT

The day and time of the visit should be fixed well in advance. This allows the adviser to arrange his other calls to allow total compliance with any unit health regulations and to give undivided attention to that unit and its problems.

The pigman can also use the arrangement of such a date to prepare:

- Results analysis—comparing current with previous levels.
- List of all treatments given since previous meeting.
- Plans for any changes in herd operation for vets' consideration.
- Arrange for any staff members who may be involved in any of the above points to be present.

It is important to note that preparation by the pigman does *not* include any 'special' cleaning or other misleading changes to unit routine in order to impress the visiting consultant! The attitude of mind must be to take the adviser into complete confidence, so that he can fully and constructively judge unit practices.

Having arranged the visit and made the preparations listed above, it is also useful to produce a plan or agenda for the day so that all points of concern are covered. Many unit personnel find the most useful plan for such a visit to be:

1. Consider comparative results and treatments—possibly with other staff members involved.

2. Walk the unit and inspect all pig and management routines and, possibly, discuss problems, or potentially problematical areas, indicated in results and treatments session. May also include debate on any proposed changes.

3. Finish with a review session and draw up an action plan; be committed to communicating the proposals arising from the meeting to all staff involved (see Chapter 20).

The time allowed for such a visit will depend upon the size of the unit and the extent to which changes are planned. Experience shows that it is difficult to conduct such a visit in comfort in under three hours, and longer may have to be programmed in many circumstances.

KEEPING DISEASE OUT

There are several important sources or potential carriers of disease organisms and these avenues for incoming problems should be barred—and, preferably, sealed.

Common, and obvious, sources of foreign organisms are:

- Other pigs.
- Vehicles.
- Personnel—neighbours, salesmen, drivers, advisers, friends.
- Other living creatures—dogs, rodents, birds.
- Weather—some organisms may be wind-borne.

Although the extent to which disease may be imported may be influenced by unit policy and design which is outside the pigman's control, once a routine is established it must be strictly followed. The vital influence which the operator has on the introduction of disease on to a unit can be illustrated by many breakdowns in Minimal Disease (S.P.F.) established herds where the cause of the change in health status can be traced back to a lapse in unit routine procedures. Practices for keeping disease out are now recommended.

Reducing the Problems Associated with Incoming Pigs

Clearly the simplest aid to reduce this problem is to purchase *no* pigs at all. This is, however, unacceptable to the fattener of pigs and to most breeders who cannot maintain a sufficiently good rate of genetic progress without the purchase of replacement gilts and boars.

In the case of incoming breeding stock, the advice is to buy in as few animals as possible, but this decision may not be taken up by the pigman whose main task is to reduce the risks associated with incoming pigs. For recommendations on the management of the bought-in weaner, see Chapter 16. The source of replacement pigs should be chosen with care and the health status of the herd supplying the stock should be scrutinised regularly to minimise the likely extent of imported disease. Once again this may well be beyond the control of the pigman.

So, it is possible to consider the fact that replacement breeding stock are to be delivered to the farm on a prearranged day. The following risks should be considered:

- The incoming stock might be incubating organisms potentially harmful to the herd, even though they show no signs of illness on arrival.
- The new animals may have had no contact with the organisms present on the farm, so may have no resistance to them.
- The incoming pigs may multiply the disease organisms present on the unit and cause a 'flare up' among the resident herd.

To reduce the likely effects of the first point, the new stock should be isolated from the main herd until the period of risk can be checked. In addition, they should be challenged with the 'bugs' from the resident herd so that they become acclimatised to the organisms on the unit and build up some immunity prior to taking their place within the main herd—thus reducing the risks of the second and third points above.

At this point it is proper to consider in a little more detail just what potential problems these animals may be bringing in with them or what organisms within our own herd they may be susceptible to. As will be seen, in the paragraphs which follow, there is a degree of conflict between the various groups of disease.

1. Respiratory diseases (enzootic pneumonia and rhinitis).

If these conditions are known to be present on the farm, care should be taken not to upset the balance of the disease by allowing incoming bugs to multiply and reinfect the main herd with a higher dose. If the incoming stock comes from a source free of these conditions, but the resident herd itself has the disease, then care must be taken to acclimatise the stock to avoid a serious dose of the organisms.

2. Scours (E. coli, swine dysentery).

Although there are many variants, consideration of these two main groups may be applied to other problems. Coliforms are present in all pigs. Thus the chance of incoming stock bringing in coli is high. Scours caused by coli are largely under the pigman's control and are referred to in the hygiene routines which follow in this chapter. The same cannot be said of swine dysentery. This condition can be eliminated and reassurance of the status of the supplying herd should be sought to ensure that the disease is not brought in.

3. Reproductive viruses (parvo-virus, enterovirus, or SMEDI).

Like the coliforms, these organisms are particularly difficult to eliminate from any stock. It is possible that, although immunity levels in the supplying herd are high and few signs of infertility ever occur, when they are introduced into another herd they could cause problems. Conversely, the resident herd may have a high level of immunity and the incoming stock might react very unfavourably to conditions on their new farm due to their low levels of immunity.

4. Other conditions: (a) Meningitis.

This is now considered to be transmitted via the carrier animal, so particular enquiries of the supplying herd should be made, as little can be achieved to prevent the entry of the organisms via isolation and acclimatisation routines.

(b) Mange.
This skin condition is transmissible, so all incoming stock should be treated prior to the transfer to the main herd.

Establishing the Acclimatisation Routine
Remember that major stresses are imposed on a pig by:

● Transport
● Change of routine
● Change of feed
● Change of housing system.

Pigs are, therefore, more susceptible to disease at these times. The idea that isolation means any kind of rough, makeshift housing, is mistaken. The quarters in which incoming stock are housed should be:

● Warm and comfortable – which implies the use of straw bedding where possible
● Separate from the main unit
● Secure
● Designed also to accommodate resident stock within the same structure to provide in-contact 'vaccination'.

Preparing the quarters

Remember that it is necessary to build up immunity in the new stock to disease-causing organisms which exist on the main unit. Therefore, *unless* there has been a serious health problem within the pens to be used, it is best:

● *Not* to pressure-clean the premises.
● To use it continually for outgoing stock, particularly all breeding animals, and to place incoming stock directly into pens which have excreta from resident pigs present.
● To place a number of cull pigs, or those due to be sold, in adjacent pens with contact through tubular or mesh divisions.

The acclimatisation routine
As indicated previously, the aims are to build up immunity (or acclimatise) the incoming stock while not putting the main herd at risk.

The best sources of possible infection or 'vaccine' are:

- Adult pigs from the main herd for infertility organisms.
- Slaughter animals for respiratory disease organisms.
- Any untreated scouring pigs for gut disorders—although moving these pigs in itself may not be desirable if they are clinically ill.

In addition, the following indirect sources should be used:

- Cleansings and other farrowing house debris for increasing immune status against infertility.
- Dung and bedding from weaner and grower pigs for gut organisms.

The pigman must distinguish between a 'lethal dose' and acclimatisation. If the new pigs are placed directly into a finishing house the level of mycoplasma within the atmosphere might cause the animals to become very sick. Whereas the in-contact procedures with a few pigs is more likely to cause a more mild, controlled reaction and allow the body defence mechanism to produce antibodies to pneumonic agents.

The converse may be, however, that the animals used as 'in-contact' neighbours for the incoming stock are not transmitting the organisms to which resistance is required. Neither is it possible to be certain that the bedding and debris thrown into their pen will be picked up or will be 'infective'.

In other words, the effects of the acclimatisation programme cannot be guaranteed, *but* it should be attempted thoroughly. At the very least it will act as a safeguard against importing disease. To increase the effectiveness of the acclimatisation routine:

- In-contact animals should be changed at least three times during the period of acclimatisation to increase the likelihood of organisms being transmitted to the new stock.
- Acclimatisation should be as long a period as can be arranged and *no less* than 28 days before pigs are moved into the herd proper.
- No 'off-colour' animals should be removed from the isolation premises until inspected by a veterinarian. Indeed, it is sound practice to seek professional advice before transferring apparently healthy stock from isolation.

Ideally, the pigs in the isolation penning will be fed and bedded without the pigman entering their pen. If this cannot be arranged separate clothing and footwear should be used and these animals tended when personnel are leaving the unit at the end of the normal working day. Separate tools should be kept to reduce infection of the main herd in the event of any adverse reaction in the incoming pigs.

Incoming breeding pigs may be served while undergoing this period of acclimatisation, providing that the pigs with which they are mated remain in isolation with them.

Before transfer into the main herd, routine worming, mange dressing and erysipelas vaccinations may be considered.

It is important that consideration be given to any known health problems in the main herd, and any major variances to this acclimatisation routine discussed with a veterinarian.

The pigman should remember:

● No 'short cut' in this routine will be acceptable.
● It is not, in itself, a foolproof guarantee of immunity in incoming pigs.
● Acclimatisation is to protect the incoming stock from the effects of disease organisms resident on the farm.
● Acclimatisation should commence as soon as the pigs arrive.
● Efforts to increase antibody status in order to offset fertility problems by feeding farrowing house debris to incoming stock should cease a few days prior to mating. This material should *never* be offered to served gilts.
● Acclimatisation should be carried out in isolation to protect the main herd.

Reducing Problems Associated with Vehicles

Because lorries which deliver feed, fuel and pigs to the farm will inevitably have had to call on other farms, they should be considered a health hazard. Of even greater potential risk may be the lorry collecting pigs.

Ideally, such vehicles will never enter the unit proper. Feed can be deposited through perimeter fences by bulk delivery vehicles and into feed stores in bags through a door which does not open to the unit itself. Loading bays should also be sited to eliminate the need for the lorry to come into the pig unit proper.

However, where farm layout does not permit this, other tactics will have to be considered. Even where unit layout does allow vehicles to serve their function around the periphery of the farm, it is essential that the pigman *never* attempts to compromise even when very busy. Just one health routine slip-up can cause a carefully nursed health status to break down.

Every unit has the right to insist of those supplying goods that:

their farm is the first one visited during the day and that the vehicle must be thoroughly washed down prior to departure.

In any case, the lorry which collects pigs from the farm should:

never have been to another farm en route, should be thoroughly cleaned and should bring no straw on to the unit. In other words, the pigman should provide a bale or bales of straw by the despatch point ready for the haulier.

If a unit layout does not permit goods to be delivered without vehicles entering the unit, the following routine may be adopted. It is essential that the proposals listed are followed during periods of disease epidemic.

- Wheels of vehicle are washed off and disinfected prior to entering the pig area.
- Gates, barriers and fences be positioned to control vehicles and personnel.
- No vehicle enters the unit without approval of the staff.
- Drivers are provided with unit overalls and footwear before entering.
- Bagged feed may be transferred outside the unit to a farm vehicle so that the lorry does not need to come into the pig area.

Similar precautions will be applied to collection lorries with two added considerations:

- The lorry driver should never be allowed to come into the unit. It is preferable to install a transfer gate through which pigs are passed to the driver for actual loading on to the lorry without assistance from the pig staff and beyond which the driver is not allowed to pass.
- A gate must be positioned in the loading area to prevent pigs, which have previously walked on to the lorry, turning round and running back on to the unit.
- If there is any doubt concerning unit security, pigs should be loaded on to a unit trailer and transferred to the collection vehicle at the farm gate.

Reducing Problems Associated with Visitors

Although the pig industry has prospered in no small measure from the willingness of farmers to share their experience and knowledge with others, the hazards from disease now make the most sensible advice in reducing this problem to be simply—do not allow non-essential people on to the farm.

This reduces visitors to the veterinarian and adviser only if drivers of vehicles and maintenance staff are excluded.

In any case, even essential visitors should:

● Be expected to declare that their 'conscience' is clear regarding contact with other pig farms.
● Be provided with overalls, hats and footwear to reduce risks of clothing 'contact' between farms.
● Be asked to wash their hands before touching the stock.

Signs should be positioned and gates should always be closed so that it is obvious to everyone that hygiene precautions are a normal part of unit operation.

Refusing entry to people well known to the pigman or owner may lay one open to charges of being antisocial. Remember, however, that to be implicated in a deterioration in herd health status would be many times worse.

Reducing the Problems Associated with Other Living Creatures

It is difficult to eliminate rodents totally from any agricultural holding because they are such a rich and obvious source of food to rats and mice.

However, it is essential that these possible conveyors of disease and potentially huge consumers of feed are controlled. If the pigman doubts his ability to maintain control over unwanted 'livestock', advice and assistance are available from:

● Local authorities
● Commercial firms who operate a contract service for pest control.

Birds have long been suspected of playing a part in the transmission of certain epidemic diseases such as foot-and-mouth disease, TGE and SVD. Even if this is difficult to prove, it is possible that they contribute to losses of feed if given free access to buildings.

Control on outdoor units is difficult, but intelligent storage of feed and careful routine operation can reduce the nuisance. These same principles should be applied to housed pigs.

Air vents should be mesh covered and naturally-ventilated buildings should be operated in a way that makes bird entry difficult. Galvanised chicken wire is long-lasting and can be simply and quickly fixed in a manner which need not interfere with unit operation. Feed spillage should always be cleared up to make the farm less attractive to birds.

Lastly, in this category—dogs. I do not believe that dogs are strictly companionable with pigs and they can be an obstacle to pig movement in some cases. Not only that, dogs eat bones and bones are a potential source of disease-bearing organisms. If a half-chewed bone is left around the pig unit where pigs have access, the risk is evident. It may be a sound rule of the unit to have no dogs around—if the owner can be persuaded!

Reducing Weather Risks
Everyone knows that no gate can be closed to the climate. However, the pigman should be aware of the risks associated with the weather and be on his guard when those conditions apply.

Firstly, there is a greater risk from certain epidemic diseases during the winter months. This is because ultra-violet rays of the sun help to control certain organisms and, during winter, there is less sunshine. Thus, during winter months the pigman should be particularly alert to such diseases as TGE, SVD and foot-and-mouth which tend to thrive in the absence of the sun's rays. Any outbreak should trigger the establishment of tight security routines for the unit.

The other set of weather circumstances of which the pigman should be particularly aware are spring and autumn conditions when there may be widely diverse temperatures recorded over a 24-hour span of time. It is possible to think of those days when it is sunny and quite still in April and October. Such conditions are normally accompanied by frosty nights, often with fog. Therefore, whenever the weather forecast indicates 'high pressure' conditions, the pigman should anticipate that the daytime will be like summer and the night like winter. If he is not alive to these circumstances and does not make adjustments to vents and/or controllers, the temperature within the pigs' lying areas may fluctuate widely. This, in turn, is the perfect setting for an outbreak of respiratory disease.

HYGIENE AS A ROUTINE

The adoption of a strict programme for keeping disease out must be accompanied by a similar attitude to reduce the build-up of disease within the unit.

Just as an acclimatisation programme will not absolutely guarantee that incoming stock will be prevented from upsetting the herd health 'balance', so it has to be admitted that the adoption of the routines suggested below cannot totally eliminate the possibility that disease will occur.

However, the best insurance available against health problems on the pig unit is to organise unit operation to permit an all-in/all-out policy with thorough cleaning possible between batches.

Apart from the determination to ensure that the routine is followed—and this commitment is required from all staff involved—the following principles are essential to the establishment of a hygiene routine.

1. Organisation of pig throughput: If the pig flow through the unit is not well regulated, it becomes very difficult to carry out a proper cleaning routine. This is another powerful reason why it is essential to programme carefully the service routine in order that the farrowing target is achieved. Then, presuming that the unit accommodation is matched to the farrowing target, this should allow for the hygiene programme to be followed.

2. All-in/all-out: Completely emptying a building allows for a much more thorough cleaning. If some pens are still occupied, the risk of re-contamination is much greater and temperature control for the animals still occupying part of that house or room is more difficult.

In addition to making cleaning easier, the all-in/all-out policy has other important advantages. If the air-space is occupied with pigs at a similar age, or weight, the climate can be adjusted to their requirement without the need to compromise for pigs of a different stage.

Furthermore, houses which are continually stocked become the ideal reservoir of infection for respiratory diseases. Older pigs may well have overcome their own 'check', but they will infect any smaller, susceptible stock placed in the same air-space. Not only are the younger pigs at greater risk due to their size and, probably, because of a compromise in the ventilation and temperature setting, but a sudden confrontation with an atmosphere heavily laden with exhaled infective material at a time when they have just changed house and, probably, diet, can be fatal.

3. Other husbandry advantages: The organisation of unit routine which allows hygiene to be planned-up also assists in allowing attention to be focused on animals in a particular section. The merits of this are discussed in later chapters. In short, it allows the pigman to concentrate his full attention on a given area and this may help to ensure that vital routine items are not ignored.

4. Repopulation of cleaned pens and houses: The absence of pigs from a building is one of the best ways of reducing the volume of disease organisms. Thus, it is useful to make the interval between de-population and re-stocking of a building as long as can be econom-ically and conveniently arranged.

The actual practice of cleaning is explained in Chapter 6. It is important, before repopulating cleaned pens, to make sure that:

- no disinfectant remains on the surfaces as this can cause skin damage to pigs;
- the surfaces are completely dry;
- all maintenance on equipment is carried out before being replaced;
- the building is brought up to, or towards, its desired operating temperature before the pigs are placed in it. This point is of greatest importance in the case of newly weaned pigs.

Further, it is important to remember that a prime purpose of the hygiene routine is to reduce the build-up of disease organisms that would occur with continuous stocking. To achieve this aim, the pigman should:

- Shut off the cleaned house during its empty period.
- Use disinfectant foot-dips when re-equipping it or moving fresh stock in to reduce transmission of disease.
- Avoid the practice of moving sick pigs back into the cleaned area with other, younger pigs as this may act as a 'trigger' point for disease.

Aside from equipment cleaning there are other essential routines which are under the control of the pigman which must be considered as a part of normal, non-varying routine.

1. Mange control

Mange is transmitted by a small mite living upon the skin of the pig. Most modern mange dressings control the mite on the pig, but the animal is prone to reinfection from eggs not yet hatched and from contact with other pigs. The mange mite finds it more comfortable to live on the pig in those areas where the body temperature is highest—this means the less accessible parts of the body where the folds of skin occur under the legs of the pig.

To overcome these problems of reinfection and access, the pigman should:

- be prepared to treat all adult stock at frequent intervals;
- if spraying, pay special attention to ears, legs and flanks;
- dip or immerse weaners in a plunge of mange dressing at weaning if a herd problem exists.

Occasional treatment of stock on a haphazard and unrecorded basis will result in unsuccessful control of mange and cause time and materials to be wasted.

2. Other skin conditions

Other conditions may be commonly observed and the pigman's role in minimising these conditions varies slightly:

(a) Prickly skin rash. This is very common in rapidly-grown pigs on straw where irritation arises on sensitive areas of the body. The pigman can help to reduce this condition by ensuring that straw is stored well and by reducing contamination with soil or dust.

(b) Staphylococcal infection. This is seen as a brownish blotch on the skin, usually of adult stock. This condition reflects suspect herd hygiene because thoroughly scrubbing the pigs using a mild antiseptic wash will normally control it. The pigman should attempt to plan a routine which allows sections of the sow house and boar pens to be vacated and washed every year to reduce reinfection.

(c) Greasy pig disease. This is thought to be of similar origin to the condition described on the adults and therefore control of the condition in the adult will virtually eliminate its occurrence in young piglets. In young pigs the skin becomes blotchy and totally discoloured with a brownish, weeping appearance. The problem can be reduced by the pigman clipping teeth, dressing wounds and providing bedding to reduce knee damage on piglets. In this way the bacteria, which may be carried by the sow, cannot easily gain entry through the wound. Again, dipping of piglets in a mange, or antiseptic, solution may help control. Short-term improvements may be obtained by the oral administration of materials rich in B group vitamins. (Conditions [B] and [C] above are correctly referred to as Exudative Dermatitis.)

3. Erysipelas
This condition is a problem in many forms. In any degree it will cause a rapid rise in the pigs' temperature and this is a major con-

tributors to significant problems such as heart damage, loss of fertility and abortion.

If he is advised to vaccinate his herd by a veterinarian, the pigman must adopt the recording and monitoring practice which ensures that no adult animal in the herd is missed. Advice on the frequency of vaccination will vary according to farm circumstances, but it is usually dependent upon the pigman for its effectiveness.

4. Worms
A variety of internal parasites may affect the herd. The stockman will be given advice upon the routines to be followed and he must then ensure:

- that the pigs are treated as specified;
- that each animal receives its share of the product;
- that the chemical used is given at the correct dose.

The pigman does not necessarily need to identify the problem, but once parasites have been diagnosed, it is he who holds the key to their control.

SECTION IV

Chapter 9

The Boar

Including checklists for:

Rearing considerations
Acclimatising the new boar
Training the young boar
Frequency of boar use
Boars on outdoor systems
Feeding the boar
Boars which refuse to work
Infertility
Artificial insemination.

Chapter 9

THOSE EXPERIENCED with livestock frequently refer to the boar as the most intelligent animal on the farm—often including the sheepdog in that assessment! Intelligence is frequently linked with sensitivity, so it could be true to say that the boar is quite the most susceptible animal to both good and bad management. This chapter attempts to set out practical guidelines for good boar management.

The boar sets out with one or two basic disadvantages. Firstly, nature decrees that the boar carries less backfat than gilts or castrates. Secondly, he is destined to live most of his life alone except for a short period several days a week!

These facts of life must be recognised by the stockman as potential sources of trouble. They make house temperature control of great importance for a start, because the boar has a lower level of fat reserves and less opportunity to lay against other pigs to offset low temperatures. Also the normal solitary life follows a period where the boar has been reared with other pigs, so the sudden change to individual penning—coupled with a change of farm—can be quite a trauma. What can be done to ensure a smooth and successful transition from a fattening pig with testicles to a stock boar siring up to 1,000 pigs a year?

REARING THE BOAR

The period from birth to selection is most commonly *not* under the control of the pigman who is to look after the boar during its working life. However, an understanding of the potential problems associated with this phase may help a better understanding of the needs of the young boar.

Virtually all boars today are sold after some screening of their performance in the period up to selection. The idea of such screening, or performance-testing, is to identify the most efficient pigs. This is in the expectation that their offspring will also be more efficient.

What does efficient mean in this context? Basically:

● that the animal has grown quicker than average;

- that it has perhaps got less backfat than average;
- that it has eaten less food than average or has utilised its feed more efficiently as a result of producing less fat.

It is no use criticising any breeder who attempts to achieve these features—if he did not he would not be producing performers which are competitive with those of other breeders. However, it can now be seen that we may still purchase our boars at the same weight, but:

- they will be getting progressively younger;
- they will be carrying less backfat.

The pigman receiving custody of a new boar must remember these facts. Indeed, the better—that is, higher-priced—boars are likely to exaggerate this problem because they will have grown even more efficiently.

While few breeders sell boars under a declared age, it is still important for the pigman to remember that there is a difference between size and maturity and that it may be increasingly necessary to hold boars on the farm for periods between delivery and their first mating.

The other aspect of the rearing phase is the method of penning the young boar. Indications are that boars reared in isolation later tend to be less inclined to serve.

Also, it has been observed that young boars reared in pens adjacent to gilts have a greater interest in mating.

So, from the commercial pigman's point of view, he would like his boars to be reared in groups in a house shared by gilts. There is one thing to be aware of here. Returning to the opening remarks in this chapter on intelligence and sensitivity, some boars respond quite badly to:

- being reared in a group;
- being separated from the group.

Those which react badly to being reared in a group may well settle down quicker when separated. However, experience shows that it is this group of boars which has the lowest sex drive, and it is to be expected that many in this category will be rejected at selection time due to their lack of masculinity.

What of those who 'pine' badly when separated and are moved to a new farm? For a start, it is necessary to attempt to counter the change of feed. Many breeders agree to supply a small sample of the diet that the boar has been fed on their farm so that it can be mixed with the feed he is now to have. The ideal pen has solid divisions between boars (to a height of 1.5 m) but vertical tubular gates or pen fronts to give good contact with sows or gilts. Next the boar should be given some company when he arrives, not thrust into some 'dark hole' and banished for a month or so in the name of isolation!

Acclimatising the New Boar

The need to protect the resident herd from possible infection by new stock and to build up antibodies to resident bugs in the new pigs has been stressed in Chapter 8. Any period of less than twenty-eight days will be unsatisfactory to achieve these two functions. It is vital to start trying to build up immunity in the incoming stock immediately it arrives or at least as soon as the boar has recovered from its transportation. The principles of this have been explained, but there are a few considerations specific to the boar.

The boar will settle down more quickly if he can communicate with pigs in an adjacent pen. It will not only encourage him to eat, but will encourage him to be interested in mating. Where a boar has been given gilts as neighbours and still refuses to eat, it is very useful to turn a young pig—say two to three months old—in the pen with the boar as a companion. This old pigman's 'dodge' has a high success rate in encouraging young boars to eat. By changing the companion every ten to fourteen days the boar does not become too attached to his younger pen-mate who can be returned to the main herd when the boar's period of acclimatisation is over. The disparity in size prevents any likelihood of the two pigs fighting.

Although adult stock, probably cull sows awaiting despatch, are useful neighbours for incoming stock, if there is a respiratory problem on the unit some finishing pigs awaiting collection may be a more appropriate source of challenge. This will allow the boar to become more 'immune' before moving into the herd proper.

If for any reason it becomes necessary for the new boar to serve before his acclimatisation programme is complete, or a pen on the main unit is ready for him, the young sow or gilt may be moved into isolation alongside him. It may even be preferred to allow the young boar to carry out his first service in the pen he has lived in for a month or so, rather than abruptly moving him to his new, permanent pen and expecting him to work straightaway in unaccustomed surroundings.

Reference has been made to the need to provide an adequate temperature control for the boar. Too often the premises used for isolation and acclimatisation are too cold for the young, rapidly-grown boar. A kenneled lying area with plenty of straw with sufficient room (up to ten square metres) for the boar to exercise with the facility to place pigs in an adjacent pen with contact through a tubular or mesh division is to be strongly commended.

TRAINING THE YOUNG BOAR

Because of their intelligence boars respond well to a stable routine. The pigman should attempt to keep to a regular daily pattern (see Chapter 1), because this helps to ensure that boars are ready and interested in serving at the appointed times each day. The young boar should become part of that routine as quickly as possible after being moved to the main herd.

The most effective use of boars comes from having them individually housed, for each service to be monitored and their frequency of use regulated. Simply to run a boar with sows makes control difficult and may lead to young boars adopting bad habits which can lead to increasing difficulties as they grow. Large outdoor units run boars in 'teams', changing them at regular intervals, but this technique can rarely be adopted with success where farrowings are to be scheduled to suit the specific provisions of the conventional, housed pig business.

Therefore, the young boar needs careful supervision to ensure that his first services go smoothly and that the boar becomes accustomed to his work. Prior to first attempting to work the boar, entering his pen for a few minutes each day at the end of normal service routine will accustom him to being handled by an operator with the smell of other boars present.

Key considerations are:

- That the boar should be no less than thirty weeks of age (seven months, and preferably older).
- That he should be well matched for size with his intended mate.
- He should work either in his own pen, or a pen with which he has been familiarised. (If a service pen is used, the young boar should be given access to it for fifteen minutes over three to four days prior to beginning to serve after the pen has been used by other boars.)
- The service pen must be inspected to reduce the chances of the boar slipping and any obstructions must be removed.
- Ideally a small sow, rather than a maiden gilt, should be used to help to train the boar with a quieter, experienced animal providing no risk of venereal transmission is anticipated.

● Because the first service is often non-fertile, it is best to plan to buy the boar well in advance so that he can be brought to work in this sequence and his first service does not have to be relied upon in order to achieve the farrowing target for the herd. Thus the use of a young sow served twice by a more mature boar and once by the young boar should help to safeguard performance as well as allowing the new boar to be trained.

With these principles in mind, the following physical routine might be followed:

1. Move a young sow to the pen where service is to take place (if separate from boar's own pen, move boar in first so that he is settled before the sow is introduced).

2. Stand in the pen with a board ready to prevent harassment of the boar by the sow or vice versa. Do *not* hurry the boar; let him work in his own time.

3. Talk gently to the boar so that he is accustomed to the human presence.

4. Do not force the boar to mount, but direct him gently to the sow's rear.

5. If the sow is well on heat she should not move around the pen too much. The stockman should help the boar by guiding the sow to stand with her head to the corner of the pen.

6. The stockman should discourage excessive nuzzling of the sow's flanks, although this is normal courtship behaviour and must be expected. If the boar mounts, he must be discouraged from 'pawing' the sow's back with his forelegs.

7. By adjusting the female's tail, attempt to let boar insert himself.

8. When the boar mounts, observe his erection closely to check that the penis is properly withdrawn from the sheath and has no abnormalities. The penis itself should never be handled.

9. Hand assistance should be given only if anal entry is likely, or if either the boar or the sow is becoming agitated or fatigued. If hand assistance is given, disposable gloves are recommended.

10. After service, allow boar to conduct 'courtship' under supervision for a few minutes, but do not allow him to remount.

11. Return boar to his pen (if relevant). In any case, after separation, the stockman should inspect the boar to check that he is sound.

12. The service should be recorded on the service register and boar card with a note clearly indicating that this constituted the first service by the boar.

Once the boar has begun to work, he should not undertake more than two actual services per week until he is one year old.

Young boars should not be used to check for oestrus in a group of gilts until they are well-experienced workers. In any case, young boars tend to have less stimulatory effect.

If a young boar does not serve first time, the above practice should be repeated every two to three days. Great patience is needed and will receive its reward in the long run by having boars which are temperamentally good with sows and humans and which have no bad habits.

FREQUENCY OF USE OF THE MATURE BOAR

There has always been a considerable debate about this topic. Evidence is conflicting about the effects of resting boars and the adoption of all-in/all-out batch farrowing systems means that boars tend to work more spasmodically than when sows are weaned at a set interval after farrowing.

What evidence is there to suggest that continual, daily matings depress performance?

The best information probably comes from the AI centres. The sampling of semen shows that there is a wide variation in sperm density between boars but the best average sperm density follows one collection per week.

Some research workers report that periods of rest of up to two weeks between serving sows gives better results, but it is difficult to evaluate these findings when they are not weighted for the effects of sow age.

As it is difficult to assess the ability of the individual boar to cope with heavy workloads, it is safest to plan that:

● The young boar serves once a week up to twelve months of age.
● The mature boar works three times a week, preferably not on consecutive days (except when fertility is being checked—see below).

These recommendations mean that the number of boars kept should be equal to the number of sows served in the average week. In this way each sow will be served three times by different boars and the boars will not be excessively overworked. The boars with lower sperm density will be balanced by those above average.

In planning to meet these recommendations, it is important to allow for:

- The odd boar which is undergoing a rest due to some ailment, or accident.
- The variation in size which prevents certain boars being used, placing a greater work-load upon others.
- The acclimatisation period for an incoming boar which may make the boar complement 'under strength' on a unit if not carefully programmed in advance.

Other useful techniques to even out the demands upon boars are:

- The wean more than once per week, so that the sows are 'on heat' over a wider spread of time.
- To cross-mate every sow so that the boar that first serves her may not have to work again for several days.
- To intersperse natural services with artificial insemination so that boar work-loads can be kept within the suggested range.

It is probably true to say that if weanings and services can be regulated so that boars serve only once a week, there is no sense in planning for any longer rest, because it will yield little response in subsequent litter size and may even reduce fertility. In fact, any boar rested longer than two weeks must be considered to be possibly infertile at his first subsequent mating.

Where sows are cross-mated a careful check on boar fertility must be kept. Once a month—preferably the first sow served after the first day of the month so the routine is understood by everyone on the unit—each boar should serve one sow two or three times without cross-mating. If any sows return the pigman will be alerted to repeat the practice in two consecutive weeks in order to ascertain the boar's fertility.

Whilst the practice of cross-mating is associated with improved litter size and, possibly, conception rates, it may also have a disadvantage. This is in the possible venereal transfer of organisms which interfere with normal sow reproduction. In herds where discharges from sows' vulvas are observed and/or an erratic return to service pattern occurs, it may be desirable to cease cross-mating and to review pen cleanliness and hygiene. Veterinary advice on resting and treatment of boars should also be considered. In herds where such a problem occurs it may even be necessary to restrict the use of new young boars to previously unmated gilts to avoid possible cross-infection.

BOARS ON OUTDOOR SYSTEMS

There is an additional set of requirements for outdoor systems where it is usual to use boars in groups which are rotated between batches of recently weaned sows and rest pens or paddocks every twenty-four to forty-eight hours.

The boars used for such a system have to have a high sex drive and ability to work unsupervised. Some boar types are better suited to this than others. Crosses incorporating some Hampshire or Duroc genetic material may be hardier and more robust and also gain extra sex drive (or libido), as a result of hybrid vigour. Cross-bred boars also appear to be able to withstand the intermittent periods of heavy work better than the pure white breeds, maintaining body condition better.

However, there may be a slight conflict in using a 'blue' boar because it is normally considered necessary to use a sow which incorporates some Saddleback or, possibly, Hampshire genes to give hardiness. This produces a weaner which is less popular with the fattener who may prefer a Large White sire to be used on the 'blue' sow.

Young boars for these systems are normally run in small groups for their first six months and a degree of service supervision is exercised during this period while they are serving, mainly, gilts. This permits the pigman to check on the individual boar's temperament and fertility before allowing him free-access serving as a part of the main boar 'team'.

In such a system it is essential to feed boars very generously during their 'rest' periods in order to maintain adequate body condition. This may be up to twice the daily levels indicated below.

Some combination of work and rest which allows boars to rest in the ratio of three days for every one day spent with the sows is common. The pigman is required to observe the actions of boars in such a system to ensure that all of them do work and that bullying and harassment do not take place.

FEEDING THE BOAR

A change of system frequently has the effect of depressing the boar's appetite, as mentioned previously. In addition, there are times when the boar loses too much condition and higher feed levels are required.

It is impossible to give rigid feed scales, but most mature boars probably require around 2.7 kg of a medium-density breeder's diet each day if working on three occasions a week. The pigman is responsible for critically scrutinising the condition of the individual boar. Any tendency to the extremes of body condition should be either acted upon or discussed with whoever makes the unit's decisions.

Where boars are reluctant to eat or require an increase in feed level to counteract body condition, an additional feed per day should be given. This has the advantage of:

- Increasing uptake by between 5 and 10 per cent per day.
- Reducing the chance of overloading the digestive system which can act as a deterrent to the pig's willingness to serve.
- It ensures that there will be no feed left to distract the sow if the boar is expected to work in his own pen.

It is important, in any case, that the pigman avoids working the boar within two hours of feeding time (see Chapter 1, 'Dry Sow House Daily Routine'). Otherwise, problems with heart attacks or vomiting may be experienced.

Most boars will choose to eat more readily if fed in a trough. If boars are expected to serve in their own pen, however, the trough may be a source of potential damage to limbs during service if it is not raised approximately 400 mm above the floor level.

Culling Boars

Modern production methods generally lead to a greater proportion of smaller sows. However, the selection of boars with low fat levels and their relatively rapid early life growth, tends to produce boars of large mature body size.

This fact, coupled with the need to maintain genetic progress means that boars are normally planned to be replaced after 18–24 months of work (up to 30 months or so in age). This coincides with him becoming too large to serve more than a small proportion of the sow herd in most cases.

Statistics show that boars have a working life expectancy closer to twelve months on average, so it is necessary to examine the probable reasons for early culling.

One is fertility. Here again it is necessary to stress the need for accurate record-keeping by the stockman, in order to detect this problem in its early stages.

A second problem is low sex drive (libido). Some boars are slow workers and even only reluctantly and occasionally willing to work. Attention should be drawn to these boars so that a decision may be taken to consider treatment or culling. A reluctant worker is a potential menace to full production because he:

- frustrates the sow, making her less co-operative and making the result of any subsequent services less predictable;
- frustrates the pigman, making his subsequent actions less likely to be accurate;
- inevitably places a greater burden upon other boars because the pigman will favour the boars willing to work and this may lead to lower productivity.

BOARS WHICH REFUSE TO WORK

A checklist for the pigman to consider is:

1. Is the sow really 'on heat'?
2. Does the sow smell of another boar?
If the sow has been penned with another boar and not properly covered, some boars will not serve her.
3. Is the pen floor slippery or uneven?
Bad floor conditions are a big 'turn off' for some boars.
4. Has the boar been recently fed?
Working too close to feeding time will deter some boars.
5. Is he used to the pen or neighbour?
Placing a boar in a strange service pen, or adjacent to a more dominant boar can act as a deterrent to normal work.
6. Has the boar been overworked?
A check on records of services should be made.
7. Is it too hot?
Boars are quite susceptible to temperatures over 24°C. If there is a prolonged spell of warm weather daily routine should be altered so that boars work before 7.30 a.m. and/or after 7.30 p.m.
8. Has the boar been stressed?
Reconsider your management routines with care and then, if this reveals no obvious cause, ask whoever makes the decisions to consult with the supplying breeder.
9. Would hormones help?
In some cases, injection with male sex hormone can stimulate activity and interest but may not result in long-term improvement.

BOARS WHICH BECOME INFERTILE

A checklist for the pigman to consider is:

1. Is it definitely this boar?
Careful recording will reveal whether a particular boar is to

blame, or he has simply been presented with 'problem' sows.
2. Is the boar well?
Minor ailments, such as tender, arthritic joints, can arise from poor acclimatisation routines.
3. Has he been over-worked?
Once again, records should be reviewed and allowances for the differences between individuals made.
4. Has the boar previously been ill?
Some conditions which give rise to raised body temperature lead to permanent sterility, others just cause temporary infertility. Bacterial infections, contaminated feed and erysipelas can be implicated in this category. Careful records of all vaccinations, treatments and loss of appetite should be included on the boar's record. A lay-off because of illness may, in itself, lead to temporary infertility.
5. Is his pen too hot?
Not only do high temperatures make a boar reluctant to work, they also reduce fertility.
6. Are the surroundings suitable?
Floor surfaces have been previously mentioned, but boars which have experienced minor injury due to badly designed gaps under doors or gates, or trapping their forelegs in horizontal divisions or gates when dismounting from a sow, sometimes undergo infertility due to trauma.

ARTIFICIAL INSEMINATION

As mentioned, AI is a useful aid to the proper regulation of the boar's work-load. It also has application in introducing new blood lines and, on some units, is even extensively used as the service method.

Greater variation in successful reproduction occurs with AI than with natural service. To improve the chance of success, the checklist below should be considered.

1. Check equipment
The catheter must be clean and, vitally important, dry. Semen should be stored at 18°C prior to use.
2. Timing
This appears to be more critical with AI than with natural service. In short: undertake thorough 'heat' detection routine using boar presence if possible; when 'heat' is observed, order semen,

inseminate six to fifteen hours after signs of oestrus are first detected and repeat six to fifteen hours later.

3. Practical tips

If possible pen the gilt or sow close to a boar to encourage her to 'stand', remove tip of spout from semen bottle and lubricate tip of catheter with a few drops of semen. Open vulva gently and insert catheter gently, taking care to incline it upwards. When resistance is felt, gently rotate catheter anti-clockwise to obtain lock in cervix and when catheter will rotate no further, the semen bottle may be connected to the end of the catheter. The end of the catheter should be turned upwards to help flow and sow's tail and catheter held together to safeguard against sow movement. If semen fails to flow out within fifteen minutes, pulse the catheter and apply only very gentle pressure to the semen bottle. There should be little runback of semen. Record this service.

4. Care of equipment

Re-usable catheters should be boiled in clean water for ten minutes after carefully washing, and be allowed to dry after used. The catheters should be allowed to dry in a clean warm spot for forty-eight hours before re-using and be cool before being re-used.

Always remember:

BE PATIENT.—Plan a comprehensive acclimatisation pro-gramme to allow the boar to settle down and become mature before commencing his 'work'. Train him with care and never hurry him.

TAKE CARE.—Keep him in good condition and record all services, treatments, etc. Match him for size with his mates and use him in familiar surroundings with non-slip floors and no obstructions which might cause damage.

BE CAREFUL.—To treat the boar with respect and never enter his pen or attempt to separate him from a sow or to move him without the use of a boar board. It is preferable that any boar movements, including service, are treated as a two-man operation.

SECTION V

Chapters 10–13

Gilt and Sow Management

Including checklists for:

Pre-selection management
Gilt selection
Gilt management between selection and service
Gilt feeding
Gilt service routine
Gilts which do not come 'on heat'
Artificial stimulation of oestrus

Causes of extended farrowing indices
Controlling the length of the suckling period
Sow culling—planned and unplanned
Abortions and lameness in sows
Conception rate and irregular returns to service
Infertility
Weaning-to-service interval
Improving livebirths
Influencing sow appetite
Reducing stillbirths
Sow condition assessments
Principles of sow feeding
Mastitis, agalactia, poor milking, metritis-vaginitis, urinary infections, difficult farrowings, savaging, haematomas, prolapses, worms, skin conditions.

Chapter 10

THE QUEST for more efficient pigs really relates to improvements in finishing house characteristics—feed conversion, gradings and growth rate. Most gilts result from such a breeding programme and these features are considered when selection takes place. This suggests that there will be a trend towards gilts becoming younger and less fat at selection time.

In fact it can be argued that the improvements that the industry demands in finishing herd performance makes management of the breeding gilt more difficult. After all, because there are performance considerations when selecting, we tend to retain for breeding the gilt with the better fattening house performance—that does not mean that it will have the best breeding features even though there is a trend for breeders to place more emphasis upon reproductive characteristics.

Two points worth considering which may help to focus the pigman's thoughts on management of the modern hybrid gilt are:

- calling an efficient finishing pig a 'selected gilt' is not enough to guarantee that she will reproduce well; and,
- because gilts' performance up to selection changes or improves, year on year, so the management of the gilts after selection should change or improve.

The major problem associated with improved genetics is that *the modern pig carries less fat on any given management regime than its predecessors did and satisfactory breeding performance depends, to a large extent, upon the gilt and sow carrying adequate backfat reserves.*

This *should* mean that the pigman, aware of this basic problem, considers gilt management not as more difficult (as suggested above), but as *different* from past methods used.

Many of the problems of sow condition, which will be referred to in later chapters, result from errors made in gilt management initially. It is common that errors in gilt feeding establish a cyclical pattern of weight losses which will be repeated right through the breeding life if preventive routines are not followed.

The period between selection and service is a much neglected

91

period of the gilt's reproductive career. The pigman should consider this just as he does the lactation period of the sow. In other words, he must attempt to produce the gilt at service time in ideal condition.

All too often replacement gilts are purchased and served a month or so later, having undergone a large change and having actually *lost* condition—which means backfat. Home-reared gilts are not immune from these kind of problems either. To rear them in a finishing house and move them to a pen close to boars will usually stimulate 'heat'. But to serve gilts within a few days of such a change where they are suddenly presented with a different diet and, possibly, poorer housing, may well mean that they are served in a declining body condition and this cannot be recommended.

The principles which should by now be emerging in the pigman's mind from the above comments on gilts are:

● she must start her first pregnancy in the best possible condition;
● if she does not do so, it will be difficult ever to put it right.

The second point also needs explanation. All too frequently it is necessary to attempt to correct gilt condition by generous feed scales during the in-pig stage. This is to be discouraged because it has several major disadvantages:

1. It implies that gilts were not in good enough condition when they were served, which may lead to disappointing performance.

2. Additional backfat gained during pregnancy will be lost more rapidly than fat gained prior to service.

3. Higher feed levels during pregnancy depress feed intake during lactation, which exaggerates the weight losses that sows and gilts undergo. This is the start of the 'sow condition treadmill', where a basic error in gilt management leads to a lifetime of excessive periods of weight loss before service, followed by a period of gain.

4. There is some evidence to show that higher feed levels in the first month of pregnancy actually reduce litter size.

GILT MANAGEMENT UP TO SERVICE

Almost half the gilts produced in the United Kingdom are reared on the farm where they are to spend their reproductive life. This gives an opportunity for the owner and his pigman to influence pre-service management to an extensive degree. Where replacement gilts are purchased there are fewer chances to affect the gilts' condition at mating—but that is even more reason for getting it right in the time that is available. The pigman can still have a large influence on the purchased gilt, even if it is only on the farm for a month or so prior to first service.

There may be a conflict between straight economic considerations and absolute levels of gilt output. Put another way, the earlier in life that a gilt farrows, the less feed and housing or labour overheads it has taken to produce that first litter. On the other hand, within reason, delayed service so that a gilt gains more maturity, may yield a larger litter. However, the debate upon the ideal weight and size of gilts at first service is somewhat academic because the overriding consideration should be:

● achieving the required farrowing and, therefore, service rate;
● which means that the precise timing of service is secondary to achieving full unit throughput and the utilisation of facilities.

Thus a realistic attitude towards gilt management is needed to make available a pool of gilts so that, in any given period, some degree of choice is available to the pigman as to which gilts he will serve and when he will serve them in order to achieve an acceptable balance, or compromise, between full use of buildings, litter output and feed use.

Thus the aim of gilt management should be continuous availability or choice of gilts ready for service.

Rearing of gilts tends to be geared to the most convenient system for a particular farm. There are two main choices:

1. To rear gilts on the same regime as other fatteners on the farm and to select them prior to normal slaughter despatch weight.

2. To separate the gilts from the main herd and rear them on a different programme aiming at a specific weight for age to attempt to achieve a balance between body size and sexual maturity.

It is right to operate the pre-selection regime according to that most convenient for the individual unit *providing* that the likely effects of, for example, growing gilts very rapidly are compensated for by effective management control between selection and service.

The major risk run by growing gilts rapidly prior to selection and by the purchase of gilts at around 90-100kg liveweight is that they are then allowed to enter a slower-growing phase in the critical period prior to service which, as stressed, will tend to reduce backfat reserves.

As a general guideline for the pigman, he should consider the following:

1. Presuming that gilts are selected or purchased one to two months prior to their likely service date, regardless of their weight or age, they must grow at the same rate in the period between selection or purchase and mating as they were growing in the period prior to selection or purchase.

2. The rate of gain may vary from farm to farm, but, in principle, the gilt should be managed and fed so that she gains no less than 5 kg per week in this pre-service phase. If gilts do not gain rapidly up to service they:

● may have insufficient bodyfat reserves at service time;
● will have to be fed more in pregnancy to compensate, which must be avoided whenever possible.

Not only does pre-service management affect the likely success of that first mating; it may also have a far-reaching influence on that animal's lifetime performance. If a gilt's condition and age are wrong at first service and too much adjustment to body condition is needed in pregnancy the probable effects are:

1. Disappointing litter size will occur.

2. Even if litter size is acceptable in the first litter a drop, instead of more normal increase, may occur in the second litter.

3. The pigs may be exceptionally sluggish in showing oestrus after their first weaning.

4. The rapid fluctuations in body condition which follow the need to adjust feed levels during pregnancy is a major contributor to a shortened breeding life, a high replacement rate and depressed herd performance which goes with this.

Bearing in mind the facts associated with the modern gilt it may be best to delay first service beyond the typical four week from delivery or selection so that the effects of immature age and low backfat reserves can be offset and a sound guide for pigmen is:

● avoid mating gilts until they weigh at least 120kg and preferably 130kg;
● aim for a minimum backfat level of $P_1 + P_3$ of 40mm or the body condition typical of such a level of fat cover;
● try to delay first mating so that gilts are at least 32 weeks of age – which is likely to be more like eight weeks from selection or purchase than four.

In terms of the lifetime benefits of getting this phase of

manangement correct, the delay in first service is not difficult to justify.

A further consideration in the rearing phase is the effect of housing gilts in proximity to boars. While close contact in the juvenile phase may help the boars to show interest, it may have the reverse effect on the gilt. Gilts appear to become so used to the presence[5] of boars that little effect from their later exposure to them results unless they can be separated for the selection–service phase.

The need to quarantine and acclimatise all incoming stock with care has been stressed in an earlier chapter. It should not be presumed that gilts reared on the farm need no acclimatisation. It is necessary to expose farm-reared gilts to adult stock, farrowing house debris, dung etc., in the period prior to service. Once again, it is not enough to deposit one piece of farrowing house debris into the selected gilts' pen or to leave the same sows alongside them for a few weeks. The chances of antibodies being built up against disease organisms will be increased by exposure to the debris from a number of farrowings and to a variety of other pigs. As with purchased stock, contact with cleansing or stillborn piglets should cease a few days before service is planned to avoid any risk of body temperature increase around service time.

SELECTION OF THE GILT

Where the pigman is responsible for the selection of replacement stock, he will consider the animals on two main grounds—performance and appearance.

Performance

As stressed, there is a need to maintain satisfactory genetic progress. Therefore, gilts which are below average weight for age may be rejected. Some assessment of backfat depths may be made, but it is not reckoned to be cost-effective to undertake ultrasonic backfat measurements on anything other than the nucleus or grandparent generation. However, where there exists a herd grading problem, there will be some merit in using ultrasonic measurements to ensure a more rapid improvement of fat depths over a relatively short period of time—say two years or so.

Size of the litter is poorly inherited so it has become accepted practice to place less emphasis upon reproductive traits when selecting gilts. However, it is wrong to discount these considerations entirely, particularly in a herd troubled with below-average litter size. Alternatively, in a herd with satisfactory finishing herd performance, it may be decided to place greater emphasis on reproductive elements, providing this can be conducted without adversely affecting other traits.

Appearance
The major considerations are:

● soundness of constitution, that is, legs, feet and teats;
● any known 'weakness' in herd conformation, such as heavy shoulders;
● ensuring that sufficient size of the gilt pool is maintained.

In other words, the finer points of visual appraisal should be of secondary importance to the above three points. If a gilt is sound, but not particularly 'beautiful', she may be retained rather than allowing the herd size to be adversely affected.

It is proper to place major emphasis on strong, straight legs with large, even-sized cleats (or cleys). Gilts should walk straight and sound and stand well up on the cleats without 'knuckling over' at the pastern joints just above the foot.

Soundness to breed dictates a need for a properly formed vulva and for at least six well-shaped, prominent teats on each side of the belly. The teats should start well forward and be evenly spaced to give adequate suckling opportunity to subsequent litters.

As far as general conformation is concerned, selection provides an opportunity for the pigman to attempt to correct any particular problems within the herd or to improve certain features. Thus emphasis may be placed on loin or ham shape, length, lightness of head or shoulder, as is most appropriate to the individual herd.

Temperament should not be ignored and records should be used to identify families or individual sows which display unwanted characteristics such as savaging or aggressiveness. Indeed, records can contribute to the elimination of other unwanted features such as genetic abnormalities—pigs with such a record will only be selected if there is a likely shortage of gilts if they were not otherwise retained.

Although the finer points of appearance and temperament should not be ignored (once the requirements of constitution and fitness to breed are satisfied), they become secondary to the maintenance of satisfactory herd size.

The required replacement numbers must be carefully planned. The herd recording system indicating likely shortfalls in sows ready to serve is set down in Chapter 4. Few herds retain sows for longer than three years on average, and in many cases it is much less. Where surplus gilts are reared there will be a cost factor. Although retaining a gilt for one extra month before discarding may cost up to twenty pounds, this is far better than incurring the loss of potential which would occur from farrowing one litter too few. That could have a cost ten times higher than that of a surplus gilt which was later rejected.

In short, the likely replacement requirement must be known well in advance so that gilt selection or purchase can be planned accordingly. Even with advance programming there will be unforeseen culling of sows and, as stressed, the only way to compensate for unpredictable shortfalls is to operate a gilt pool approach. Thus gilt selection should be geared to provide for up to a 10 per cent surplus of gilts being available.

FEEDING THE GILTS

The common errors in gilt feeding and their effects upon lifetime performance have been referred to.

The results of typical feeding and housing management are shown in fig. 5, where feed levels have not compensated for:

● change in farm or housing;
● change in diet;
● change in routine;
● quality of environment.

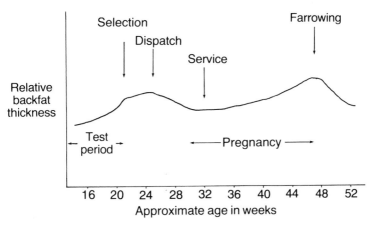

Fig. 5. Effects of typical gilt feed pattern on backfat reserves

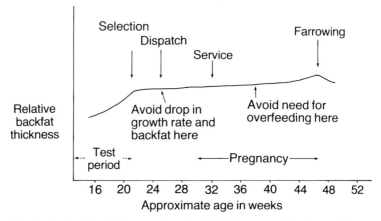

Fig. 6. Desirable gilt feed pattern to initiate improved body condition

Thus the gilt is quite probably served in a declining state of fitness and a lower state of fatness than when selected or purchased.

This leads to the later rush to compensate which leads to the cyclical loss of condition indicated.

The aim must be to follow the pattern emphasised previously and aim at the desirable body condition effects shown in fig. 6. This also has a similar effect upon stimulating full ovulation as 'flushing' the gilt with extra feed prior to service.

The absolute levels of feed required to achieve this pattern will vary according to:

● rate of growth previously achieved up to selection or delivery;
● quality of housing;
● quality of diet;
● whether gilts suffer any health check during the acclimatisation programme.

It is proper, in any case, for the pigman *not* to have a too rigid opinion upon the feed scale, but to have sufficient flexibility to achieve the recommendations shown in fig. 6.

It may be helpful to consider the following general suggestion:

1. Feed generously up to service *so that . . .*
2. there is no need to feed generously afterwards.
3. It is not feed used which really matters, but the output from that feed, so if a little extra feed produces better results it can be more than justified.

SERVICE ROUTINE

The operation of a gilt pool approach implies that gilts will be housed in groups. It may be of some advantage to use a group housing system for gilts at least beyond service time because group housing:

- encourages onset of oestrus through excitement created by contact;
- encourages the individual gilt to display more obvious signs when her 'heat' periods are normally shorter or less obvious than those of other gilts;
- allows extra movement which may help gilts to settle more quickly into new surroundings and reduce the effects of stress associated with such a move, particularly the loss of appetite.

At selection time it is usual to give each gilt an erysipelas vaccination. This should be recorded by the pigman. It may be desirable to give a booster vaccination when the gilts are transferred to the main breeding unit.

Once gilts have been selected, or purchased, it is probably an aid to the pigman to identify the gilts with an easy-to-read plastic ear tag. This enables any treatments given, or oestrus displayed, to be recorded to assist with future planning.

When gilts are required to come on heat, stimulation can be achieved by placing an active boar in their pen for a short period each day. This should be done under supervision to ensure that:

- harassment does not occur which may deter the gilt rather than encourage oestrus;
- unplanned services do not take place.

This routine will be more effective:

- where an experienced boar, which exudes greater boar odour, is used (the boar should not be so large that he frightens the gilts);
- different boars are used in rotation.

This practice should be used to trigger oestrus and a note made when gilts show signs of 'heat'. It should be discontinued when the gilts come on 'heat' to avoid indiscriminate mating.

It is more conducive to satisfactory litter size to ignore the first oestrus displayed by the gilt and to serve at the second or third period, providing that this fits in with the service programme.

For service, gilts thought to be 'on heat' should be moved into the pen in which service is planned with a boar of suitable size and

Plate 5. The close proximity of sows and gilts to boars aids onset of oestrus, makes detection easier and helps to keep boars active.

careful observation of the courtship and service made. No gilt should be left with a boar to be served because their excitement and inexperience frequently lead to them standing in a position where it becomes difficult for the boar to mount and serve properly.

The service will, ideally, be repeated twice, so that the gilts are treble-served during a 24-hour period (see details under sow service, Chapter 11).

It is essential to make a careful record of service details so that likely return dates can be checked.

Gilts Which Do Not Come 'On Heat'

A checklist for the pigman to consider is:

Is it the whole group of gilts which show no signs?
If **Yes**:

Are the gilts in good enough condition? If not, feed more feed, or a better diet.

Are the gilts too fat? Unlikely in the modern gilt, but withdrawing feed for twenty-four hours may stimulate oestrus through stress and excitement.

Were they reared too close to boars? Introduce different active, older 'smelly' boars to them, or use synthetic boar odour spray.

Would a change of pen help? Moving pigs into a pen previously occupied by a boar or sows 'on heat' can trigger oestrus.

Is there enough light? Moving gilts into a darker pen may depress oestrus—use artificial light for sixteen hours a day.

Do they seem too docile? Run an active, not too large sow that is well 'on heat' in with the gilts, so that she creates activity and excitement. Repeat with a different sow for several days and remember that gilts housed in groups tend to display quicker, stronger signs of oestrus.

Could diet be having an effect? A poorly stored diet can reduce vitamin and mineral availability. High fat/oil inclusions must be balanced with extra supplementation of vitamin E and selenium—check with whoever makes the decisions.

Is the feed contaminated? Raw materials contaminated with mycotoxins leave a reservoir of contamination for future consignments of feed through bins or feeding equipment. Equipment should be cleaned regularly and bins and stores coated with mould inhibitors at least twice per year.

Would extra vitamin injections help? Decision-maker may seek opinion of veterinarian and nutritionist on additional administration of, in particular, vitamin E injection.

Are the gilts healthy? Sub-clinical illness may have a depressing effect on the onset of oestrus. Slightly reduced appetite, reluctance to move, change in dung texture, slight temperature may not be easy to detect, but could all affect oestrus adversely.

Are the gilts too young? There is a breed influence in determining sexual maturity and this may be exaggerated by very rapid growth in the pre-selection phase (see previous notes in this chapter).

Is it the individual gilts which show no signs?

Is the individual being bullied by her pen-mates? Just as some boars respond badly to more dominant pen-mates, so the timid gilt may not show oestrus so obviously.

Is the individual normal and healthy? The comments above on sub-clinical conditions may depress oestrus in the individual. If gilts are allowed to cycle too many times before being mated, there appears to be a greater chance of their ovaries becoming cystic. Records will help to determine the likelihood of this occurring.

Is it a consistent problem?
If **Yes:**

In addition to the above considerations, there is a need to review whole gilt management routines.

Consider the acclimatisation programme and in particular whether it is possible that there is a disease influence. Such problems are best discussed with the unit veterinarian as, indeed, any supplementary administration of vitamins should be. It may be necessary to resort to the artificial stimulation of oestrus by means of hormone injection as a short-term measure to enable a normal service programme to be achieved.

A possible tool in herds with a persistent problem, or in a newly formed herd, is to use a vasectomised boar to run with the gilts in an effort to stimulate oestrus.

If the problem recurs on a unit then there must be some antagonistic influence between environment, feed and health. If that is *not* the case then doubts must be expressed about the capability of the operator to actually observe or to carry out the routines detailed.

Any changes to management which are made, and any treatments given, must be carefully recorded so that the pigman and his adviser can consider the effects of these alterations.

Artificial stimulus of oestrus
Whilst the preference is always to seek veterinary intervention only as a last resort, there may be occasions when treatment may be considered.

Synchronising oestrus If establishing a new herd or substantially expanding an existing one there may be an advantage from accurate synchronisation of matings and, therefore, farrowings. Progestogen added to the gilts' daily feed allocation for a 19-day period following earlier observed oestrus and then suddenly withdrawn can have the benefit of a group of gilts displaying oestrus more or less simultaneously. This allows boars or AI to be arranged to permit mating without over-use. Care to ensure that all gilts receive the proper quota of the drug is vital.

Failure to show 'heat' Two main groups of drugs exist but their effectiveness will depend upon the reason for no 'heat'. If the gilt is truly anoestrus one group of drugs will be ineffective. If the gilt is merely not displaying 'heat' but is cycling normally ('silent heat') then the other group of drugs would be more appropriate.

The problem for the pigman and his veterinarian is to guess which of the problems exists as use of the wrong drug will waste money and may even disrupt the gilt's natural hormone cycle with serious long-term effects. For this reason such products should be used only after careful consideration of all the husbandry techniques described.

Always remember:
Feed gilts well up to service, and *less* thereafter.
Subject them to contact with dung, debris and older breeding animals even if they have been reared on the same farm; *to boar contact* for three to four weeks prior to service time and to sows 'on heat' if the gilts are slow to react.
Consider a delay in mating to ensure that body condition and age are suited to a long and fertile lifetime performance.

Chapter 11

At present there is little prospect of being able to shorten the length of gestation safely by more than one or two days. However, there is a real prospect of being able to influence significantly the duration between farrowing and conception by the application of good management.

The pigman may not control the choice of facilities which allow a particular weaning age to be practised. Weaning age, itself, plays a major part in the rebreeding interval. Nonetheless it is possible, by ensuring close adherence to the theoretical weaning age for a particular unit, to ensure that full use of the unit facilities is made and that no 'slippage' in lactation length occurs. The pigman is responsible for eliminating anything which might cause weaning to be delayed beyond the planned stage.

Theoretical farrowing indices

Components of length of reproductive cycle	Average age at weaning (days)			
	14	21	28	35
Pregnancy	115	115	115	115
Lactation	14	21	28	35
Weaning–service interval	9	8	7	7
Theoretical days/ cycle	138	144	150	157
Theoretical farrowing index	2.64	2.53	2.43	2.32

	Farr. index	Days lost	Farr. index	Days lost	Farr. index	Days lost	Farr. index	Days lost
Effect of some	2.6	2+	2.5	2	2.4	2	2.3	2
typical farrowing	2.55	5	2.45	5	2.35	5	2.25	5
indices upon	2.5	6	2.4	8	2.3	9	2.2	9
length of	2.45	11	2.35	11	2.25	12	2.15	13
reproductive cycle	2.4	14	2.3	15	2.2	16	2.1	17

Very few herds are more than 90 per cent successful in achieving the potential for frequency of farrowing. The table illustrates the theoretical capacity for four different weaning ages. It also shows how the achievement of what are accepted as satisfactory indices of farrowings represents a considerable shortfall from the calculated levels of performance.

WHAT CAUSES DAYS TO BE LOST

Common causes of variance from the theoretical level of litter output are to be found in the checklist below. It must be stressed that it is unreasonable to anticipate that every sow will 'hold' to her first services. Nevertheless, there do exist certain individual herds where this does not occur more frequently than in one or two per cent of all services. However, this should always be considered as a possible area for investigation. It is stressed that:

5 per cent returns to service add an extra day, on average, to each reproductive cycle of every sow in the herd.

Next, the results themselves should be scrutinised closely to ensure that the correct interpretation of the results is being made. In other words, before considering the checklist below, it is wise to make sure that changes in the herd structure itself are not masking the true state of affairs in the herd.

Commonest causes of confusion are the change of herd size, or the way that animals are classified. For example, increasing the herd size produces effects not seen in a herd of near static sow numbers. The introduction of a number of served gilts into the herd:

● increases the number of animals which are 'divided' into the number of farrowings and shows that litter output appears to have fallen.

By recalculating herd size and removing those gilts which give rise to the increase in herd size, a more accurate calculation could be made.

Another manner in which changes in the herd might affect the appearance of results is in the proportion of gilts in the herd. There are two influences here. Firstly, if gilts are considered as sows at the point at which they are transferred or purchased into the breeding herd, they will have the effect of increasing average length of the reproductive cycle. Conversely, if they are transferred in at the point of service, or even conception, this will tend to make the output appear better.

The second gilt influence is on the average herd age. Regardless of the point at which gilts enter the breeding herd, an increase in the proportion of gilts within a herd of static size will affect the average length of the reproductive cycle. This is particularly true where gilts at the point of service are taken into account in the calculation for litters per sow per year. If this point is considered as the stage at which they replace a sow, it can have two effects in the herd of static size:

1. It appears to bring about an improvement through eliminating the weaning-to-service interval which would apply to sows if *not* replaced.

2. If gilts are served in advance of the weaning of those sows they are to replace, this can have the effect of making results appear better than they actually are. The converse is also true.

Efforts to allow a more realistic appraisal of results include using a system of results analysis which 'lags' the age of the herd. Simply, this means that instead of calculating litters per sow per year on the average herd size over the last twelve months, the average herd size over the twelve months up to the point when gilts and sows were served is used. The table (page 107) illustrates the effect of this in a herd increasing from 200 to 250 sows by inclusion of additional gilts.

Once it has been established with reasonable certainty that the herd statistics are not concealing the true state of affairs, the pigman may give consideration to the following checklist.

1. Is the suckling period actually of the length estimated?
2. Are sows being retained for a long interval after weaning before being culled?
3. Are there excessive unplanned cullings of served sows?
4. Is the conception rate too low?
5. Are there too many sows served, but not farrowing?
6. Is the average weaning-to-service interval too great?
7. Are sows returning at an irregular point following first service?

Ensuring Accurate Control of Suckling Period

If sufficient building facilities exist for the size of herd kept, it is essential that throughput is closely and accurately maintained. This means that:

● herd size must be precisely controlled;
● service targets must be accurately maintained;
● records to ensure that the first two points are achieved are

Effects of change in herd size on calculated litter output

Today →

Extra gilts served here

Actual herd size (includes in-pig gilts)	200	200	200	200	200	200	200	200	200	210	220	230	240	250	250	250
Average herd size (actual size/month ÷ 12)	200	200	200	200	200	200	200	201	202	208	213	221	225	224		
Herd size from which farrowings occur.	200	200	200	200	200	200	200	200	200	201	202	208	200	200	202	208
Period used for calculation, i.e. 'lagged' period.																

4-month pregnancy span

N.B. The effects are reversed if herd size falls, i.e. will make results appear worse due to fewer farrowings occurring relative to the average herd size.

Note that the gilts brought into the herd in the shaded box on the top line do not farrow for a further six months, thus distorting the calculated output per sow.

Average herd size here will give 'false' litters/sow calculation. Calculated litters/sow/year i.e. = 448 ÷ 229 = 1.96. but true average = 448 ÷ 208 = 2.15

essential, as is knowledge of the actual days from farrowing to weaning.

In other words, the pigman must ensure that bad organisation and bad record keeping or interpretation are not influencing results. Apart from the failure to predict accurately and control weaning age, the likely reasons for lactation length being extended beyond the predicted time are:

1. too slow post-weaning growth leading to extended period of time on-farm and insufficient space relative to calculated needs;
2. excessive number of pigs too small at weaning to thrive in existing weaner accommodation, giving rise to:

● extended suckling period;
● excessive fostering of small piglets on to other sows.

The means of overcoming such difficulties are discussed in detail in subsequent chapters. However, you should check that the weaner pig housing is adequate for the age of pigs to be weaned and discuss this with whoever makes the unit decisions.

Sows Being Held on the Farm Too Long Before Culling

An important, and neglected, area of planning is the stage of sow removal from the herd. Once the decision to cull a sow has been taken, it rarely pays to retain her in order to gain extra weight. Thus, once her udder has returned to normal and despatch is convenient, she is best removed from the herd to make way for a more reproductive sow.

Where removal is delayed, the effect on the length of the average reproductive cycle can be considerable. In a herd where one-third of the sows are culled each year, a delay of one week longer on average than necessary in disposing of each sow will add one day to the average reproductive cycle of each sow.

Excessive Unplanned Culling of Served Sows

Very few of the reasons for removing sows from a herd can really be said to be completely predictable in advance. We may know that an estimated number of replacement gilts would be required, but it is less easy to forecast the exact timing of such a requirement.

Figure 7 shows the results of a survey published by the Meat and Livestock Commission in 1980. Only the culling of sows by age can really be said to be predictable in advance.

The proportion of sows culled after service totals at least 47 per

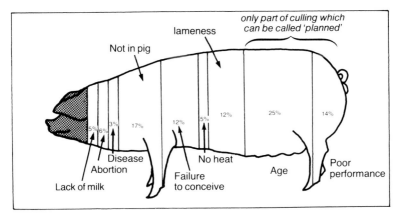

Fig. 7. The planned and unplanned culling of sows

cent of the reasons for removal from the herd. Not in-pig (17 per cent, see below), failure to conceive (12 per cent, see below), abortion (6 per cent) and lameness (12 per cent) are discussed under this heading.

Lameness constitutes a major cost on replacement rate. It will, in the main, reflect the facilities on the farm. The pigman's role in preventing lameness should be:

1. To review all his actions to ensure that movement and handling procedures are not aggravating the condition.

2. To reconsider acclimatisation procedures to ensure that infection is not entering the joints as a secondary invader to, say, mycoplasma challenge.

3. To use records to indicate whether a particular pen or group of pens predisposes animals to these conditions.

4. To ensure that pens are as dry and draught-free as possible.

5. To check that damaged kerbs, slat edges and floor surfaces are reported immediately for maintenance attention.

6. To carry out as recommended any treatment programme advised by a veterinarian.

A useful treatment for, or prevention of, lameness is to walk sows regularly through a footbath containing 5 per cent formalin or copper sulphate inclusion. Regular passage through a footbath (at least twice-weekly), not only assists the control of infection, but has the effect of hardening the hoof.

Abortions can be of occasional significance on a breeding unit. It must be considered as being abnormal when:

abortions exceed 2 per cent of the served section of the herd.

Once a sow has been served longer than thirty-five days any major trauma may induce her to abort. A major cause will be any factor which causes a significant rise in body temperature. This may include:

● a local infection, such as kidney problems, wounds, erysipelas;
● stress or physical damage—fighting, broken floors or slats may cause the sow to be distressed. Sows should also be fed in the same sequence to avoid extra excitement.

Feeding may also play a part in the abortion problem. Frequently this is related to poor food storage or bulk bin management, but accidental contamination of feed or a sudden change in diet can trigger abortion. Where feed stocks become contaminated with mycotoxins, or mouldy feed, the problems may occur (see Chapter 3 for bulk bin management).

Only when the above possible causes of abortions have been thoroughly cleared of any implication should reproductive organisms (SMEDI) be implicated and veterinary advice on that problem sought.

Standards for Sow Culling

Hard-and-fast guidelines for planning to cull a sow cannot be given. In any case the pigman is advised firstly to *ensure that sufficient replacement gilts are available for service to compensate for unforeseen sow disposals.* In other words, *plan* for the worst eventuality, do not *hope* for the best!

There are two standards for culling which can be pre-set in advance: sow age/number of litters produced and piglets produced. Other reasons for culling cannot be predicted, but should be considered on the basis of being within a predictable proportion.

Age of sow
Figure 8 shows a typical distribution of litter output in a commercial herd. It should be noted that sows reach a peak of litter size around the fourth of fifth farrowing and, in particular, live births decline after this time. The pigman should reflect that:

● After litter six live births decline in relation to stillbirths.
● Sows with a good reproductive record may still out-perform gilts.
● Older sows will have a higher tolerance of, or immunity to, the disease burden on the unit.

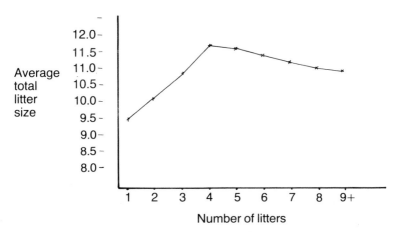

Fig. 8. Typical distribution of litter output

● That it is better to keep an older sow than have insufficient numbers of sows farrow.
● Any delay in culling sows must be known in advance so that a sudden decline in the herd size at a later stage does not upset herd output or jeopardise the health status following a larger influx of gilts.

If herd age is varied due to a change in policy or poor planning, a significant impact upon litter output may result—as shown in the figs. 9 and 10. Notice how a change in average herd age, at standard levels of litter size per parity, can depress the herd average for numbers of pigs born.

Clearly, if average age of sow is allowed to become too high it will be followed by a period when it is too low and unbalanced in the opposite way. Herd statistics should be studied so that the aim can be set for an average length of life of sows which is realistic and achievable. It is especially important to consider the average age of sow in a herd established in the last four years, or which has undergone a substantial increase in that time. If replacements and culling are not carefully matched, it is possible that too many sows will be removed at a similar time, giving rise to peaks and troughs in the average age of the sow herd on a three-yearly basis.

Reproductive performance
Sows may be culled due to their failure to produce sufficient pigs, to rear sufficient pigs, or to reproduce regularly enough. The setting of

	Parity	1	2	3	4	5	6-8
Target		22	19	17	15	14	13

Fig. 9 Suggested parity spread for a herd of constant size

standards for these factors cannot be absolute and rigid within a herd. A useful guideline is *to compare the performance of the individual sow with the herd average*. Thus, ignoring the first-litter sows, those which fail to give birth and rear as many pigs as the herd average with as good a farrowing rate as the herd average should have their future carefully reviewed. Again, 'doubtful' sows as far as performance is concerned, would be retained rather than lead to a shortfall in herd size.

Records should be consulted to ensure that there are no 'hidden' problems. For example, a sow may appear to be rearing large litters, but if this is because she is being used to foster-rear piglets from other sows, her presence should be reviewed.

Simply, if the average length of the reproductive cycle is 150 days on the unit, any sow more than ten days outside this over two litters or more should be a candidate for culling.

The average number of pigs born and reared should be considered against the herd average, although some may prefer an arbitrary standard regardless of the herd mean. If a unilateral standard is to be used the following standards might be considered. Sows will be considered for culling where:

They rear less than 17 *pigs in their first* two *litters*
28   three
39   four
50   five
10 pigs per parity, on average, thereafter.

In addition to these three main performance considerations, a herd with a satisfactory level of sow output may switch emphasis to other important performance elements. These might include:

● *Birthweight.* This is a vital element in piglet survival, so sows which produce total litter weights below 15 kg or produce more than 10 per cent of pigs born under 1 kg might be considered to be below average.
● *Milking ability.* In a well-managed herd this should be detected by low numbers reared because a sow failing to milk sufficiently

Parity	Effect of actual parity spread (%)	Average litter size for each parity	Effect of target parity spread* (%)
1	29	9.38	22
2	22	9.79	19
3	26	10.73	17
4	12	12.21	15
5	9	11.95	14
6-8	11	11.46	13
Herd Average Numbers Pigs Born	10.39		10.89
Difference in average herd litter size = 0.5 pig per litter			

*See Figure 9.

Fig. 10 Impact of herd age/parity distribution on average litter output

to rear an even-sized litter of good pigs should have pigs fostered off. However, where this is not or cannot be practised, those sows which produce below-average total weaning weights, or where more than 10 per cent of their piglets are more than 10 per cent lighter than average may have a question mark placed against their milking capacity.

Other considerations
Clearly those features which contribute to poor performance but which the pigman himself might work at in order to reduce the effects on output should be recorded.

Good examples of such features are protracted and difficult farrowings and savaging of piglets. If the correct farrowing routines are followed, the effects of these problems could be reduced. However, such sows should be removed from the herd if the problems persist because an unattended farrowing with these problems could result in large losses.

Remember
Culling should be planned to give a properly regulated herd age and standards adopted so that pigs of below-average performance or

difficult temperament are removed. However, culling must never be considered independently of gilt replacement availability. The attraction of increasing sow age to help increase the immunity status may be balanced by the disadvantages which older sows bring.

CONCEPTION RATE TOO LOW

The pigman should expect that *over 90 per cent of sows and gilts served should not return to service at three weeks*. Where returns at three weeks exceed 10 per cent, the following checklist should be considered:

Are the services properly supervised?
As mentioned under boar and gilt management, the pigman must ensure that proper contact and service takes place. The whole service routine must be reviewed and considered a potential problem area if conception rate drops below 90 per cent. Conception rates are always poorer where there is a poor service area floor.

Is the timing of the service correct?
Figure 11 illustrates that if sows or gilts are served too soon after the onset of oestrus ovulation may occur too late for conception to take place. Thus a service routine which allows sows to be served three times when 'on heat' will help to mask this effect.

Are the boars fertile?
This is discussed in detail in Chapter 9, but also includes the regularity of boar use which is under the pigman's control. Boars used too frequently will have a lower level of fertility.

Are the sows in suitable condition?
A big influence on conception rate is the change in body condition that the sow undergoes in lactation. If she is weaned too thin, or has lost excessive condition while suckling, her chances of conception after weaning are reduced. Sow feeding details are given in Chapter 13.

Did the gilt or sow bleed at service?
Blood has a spermicidal effect, so bleeding at service time is almost certain to cause sows to return to service. Up to 10 per cent of gilts may bleed due to quite normal physical circumstances. Some sows which have had a difficult farrowing develop blood- and fluid-filled

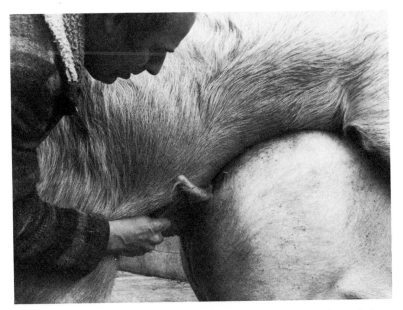

Plate 6. A check of every mating is necessary to ensure that the technique is not a cause of poor conception.

'blisters' (haematomas) on the cervix which rupture when served. Both these conditions are normal and should be anticipated at the low levels indicated.

What time of year is it?
There appears to be a seasonal decline in fertility when daylight length begins to shorten in August and September. Extending daylight effect by the provision of artificial lighting (30 lux at 1 m above floor level in the sow's area), for up to three hours per day at the time of year indicated might help.

Are the sows stressed?
Particularly where hyper-active and aggressive sows are grouped with, or even penned next to, a submissive, timid sow, the stress caused may adversely affect the conception of the submissive sow.

Are the sows affected by disease?
Likely health influences may be P.R.R.S., parvo-virus and Aujeszky's disease, but the pigman should not use these causes as excuses until

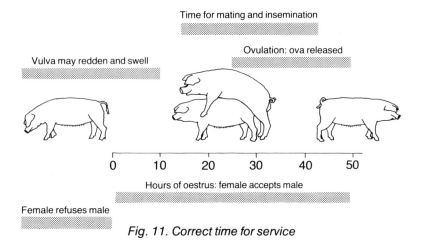

Fig. 11. Correct time for service

he is satisfied that all the husbandry effects have been satisfactorily eliminated.

Sows Which Do Not Farrow

It is possible that a distorted impression of the reproductive rate may arise from a proportion of sows which are found not to be pregnant at a late stage of pregnancy or even at the time when the service register shows that she should farrow.

The number of sows which come into this category and are not seen to abort *should not exceed 2 per cent of all sows and gilts served.*

Possible causes of this phenomenon include those which may lead to abortion and are referred to above.

However, a more likely contribution may come from stockman routine. The chances of returns to service arising are increased when:

● sows are not placed in a pen with a boar for a physical check at least once around three and six weeks after first service;
● a daily check in the dry sow area using an active boar at the gate of the sows pen or at the head end of a stall standing is not conducted;
● sows are moved away from a boar's presence into a separate house after first service.

There are occasions when a sow's ovaries become cystic which prevents her 'cycling' in the normal manner. In addition, but rarely, it is possible for a partial reabsorption of the womb contents to occur, but for certain parts to be retained. The hormone mechanism reacts as if the sow is pregnant, but she is really in a state of suspended reproductive activity.

As stated, there may well be husbandry influences upon the sow to cause reabsorption of the litter and trigger a gradual return of the oestrous cycle at a later and unpredictable time. Where there is evidence to show that this is happening in spite of good attention to routine checking procedures, using a boar, it may be advisable to consider *the use of a pregnancy detector*. Pregnancy detection is a useful aid where:

● there is a suspended fertility problem;
● it is difficult to apply physical boar and sow checks;
● there is a need for the skilled pigman to check relatively inexperienced operators.

While it is possible to undertake laboratory analyses to check hormone levels in the sow's blood, this is a time-consuming and difficult task. Despite its accuracy, the vaginal biopsy method relies heavily on the skills of a practised operator and there is a delay while samples are sent away for laboratory analysis.

During the 1970s two electronic systems of pregnancy detection were refined so that, with cursory training, a pigman could use them with good results.

The most common type is the ultrasonic device which transmits sound waves to be reflected from fluid-filled areas—like a pregnant womb. Providing that the signal is not confused with a reflection from a full bladder these machines can be used with good accuracy between thirty and fifty-five days after initial service. Pregnancy is indicated quickly by audio and visual signal.

In herds where there is a known problem with late returns or, indeed, 'full-term' non-pregnancies, the Doeppler-type machine may have the advantage of being accurate after the later limit of the ultrasonic types. These machines reflect sound signals from moving surfaces such as the blood supply to the womb or even the embryo's heart pulses. Although this machine is slower and more tiring to use it has the advantage of greater accuracy than the ultrasonic types.

Whatever machine is used, the following routine is suggested:

1. Use the machine at the same time each week.

2. Check and record all animals between four and eight weeks after service.
3. Clearly mark any 'doubtfuls'.
4. Move any sows shown to be 'negative' adjacent to a boar and check daily for oestrus. If no sign of oestrus shows in twenty-one days, the animal may be culled.

Remember
Pregnancy detectors are less efficient than a boar, but can help to reduce 'empty days' from 'problem' sows.

WEANING-TO-SERVICE INTERVAL

This is the stage which ranks with farrowing time as needing maximum attention from a skilled and devoted operator in the breeding herd. The pigman's influence on the length of rebreeding interval and conception rate is considerable.

Within this category is the condition in which sows appear to show no sign of 'heat' at all. This is often termed anoestrus. The reasons for this condition are unclear and while drug treatment can help to rectify the problem, you should first consider those points discussed below before consulting with a veterinarian on the use of drugs. One operational technique which might be useful is an injection of vitamin E at weaning as a stimulus to the normal hormone response of the sow at that time.

It is important that the pigman knows:

● the average weaning-to-service interval;
● the range between individual sows;
● the proportion of animals which take longer than seven days to show 'heat';
● whether there is any difference between parities.

Given this breakdown of the records, it can be decided if there exists a problem with individual sows or the whole herd, or if it is a 'first litter weaning' problem. The following checklist can then be applied to the problem.

Are the sows housed close to the boar?
While the pigman cannot be expected to alter the buildings, if the pen layout which he uses prevents newly-weaned sows being penned so that they can see, smell and hear a boar, he should:

- use barriers so that a boar can be allowed to patrol the passage-way by the sows each day;
- move the sow to a boar every day.

The presence of a boar near to newly-weaned sows can shorten the weaning-to-service interval by several days on average, and may even induce better ovulation leading to improved conception and bigger litters compared to sows isolated from boars at this time.

Are the sows penned individually?
There are increasing reports that sows penned in individual pens have a slower onset of oestrus after weaning than those placed in groups. While it is not possible for the pigman to change the entire system within a unit, he may choose to use one pen, perhaps a surplus boar pen, in which he can:

- group sows which have taken longer than seven days without sign of oestrus;
- place gilts after their first litter in order to stimulate onset of oestrus.

Are the sows 'starved' after weaning?
The practice of feed deprivation at weaning is now thoroughly discredited as it has been shown that feed withdrawal:

- does *not* speed up drying up of the udder;
- extends the weaning-to-service interval.

Sows should be fed up to 4 kg of feed per day to speed onset of oestrus and to encourage full ovulation and a rapid replacement of lost bodyfat.

Is the environment satisfactory?
The importance of satisfactory temperature maintenance has been stressed and the pigman's part in this is to carry out those checks recommended in earlier chapters.

In addition to temperature control, it is thought that inadequate lighting can suppress reproductive activity; a lighting pattern which provides up to fifteen illuminated hours per day may yield some benefit in reducing the problem in some herds.

Is sow condition satisfactory?
Sows which are weaned in poor condition or those which appear 'fit' but have undergone a large relative change during the suckling

period will tend to have a slower recovery and onset of 'heat'. This may be a particular problem with gilts following their first litter and typically happens when they have eaten inadequately during lactation which, in turn, results from too heavy feeding in pregnancy (see Chapter 10).

Are the sows weaned very early?

Results show that the average weaning-to-service interval is extended with early ages of weaning even though the farrowing-to-service interval will still be short. In most cases, it is *not* a general extension of interval which occurs, but there may be more individuals which take longer to show signs.

The pigman should ensure that:

● early-weaned sows are fed generously in lactation;
● gilts, in particular, are encouraged to eat to minimise weight change between farrowing and re-mating.

Due to the time taken for the womb to recover from farrowing, reproductive activity within eighteen days of farrowing may be reduced even with diligent efforts on the part of the pigman. Some slight increase in the re-breeding interval might have to be accepted with earlier weaning.

Is service routine satisfactory?

Wherever reproductive problems arise it is necessary to review, in detail, the day-to-day management routines, ensuring that the sow is placed with the boar and her contact supervised to ensure that she is definitely not cycling or simply refusing to accept a particular boar. As stressed, it will be insufficient simply to inspect sows in their own pen to discern oestrus and it will be almost equally inaccurate to place a sow in with a boar and not supervise the courtship behaviour.

Are you sure the sow is not 'on heat'?

There are circumstances when a sow may reject a boar even though there are other outward signs of oestrus. The pigman should check carefully that:

● the sow is not simply reacting against that particular boar. There is a small element of incompatibility between individual sows and boars. This effect is heightened when a sow is penned close to one boar, but is placed with another for service. Occasionally such a sow will only accept the boar penned close to her.

● The sow is not injured in any way. Slight back or limb damage or sun scald will make the sow reluctant to let a boar mount.

Electronic aids to assist pigmen in determining the optimum time of service have been produced, but results vary considerably from farm to farm. Some users of artificial insemination claim improved conception as well as litter size when using the device and the equipment would be considered where other changes fail to yield improvement.

Are sows returning at an irregular point following first service?
As opposed to the problem of failure to farrow following a presumed effective service this problem has a quite depressing effect on results. Returns to oestrus outside a typical three or even six week cycle from first mating are doubly difficult to spot as the stockman normally pays less attention to oestrus outside these periods. Although the points discussed under the section on failures to farrow are relevant there are other considerations.

Is the return rate typically around thirty days following service? If this 'slipped' pattern is observed it is, in effect, a short-term abortion most probably caused by an infection of the uterus. Such an infection may arise from farrowing time when care, particularly assistance at farrowing, will help to avoid the problem. The infection may also be caused by sows lying in a wet, fouled area and ascending bacterial infection taking place. Thirdly, of course, the infection may be transmitted by the boars. Resolution of the last two possibilities is obvious – to keep sow and boar pens clean and to consult with the unit veterinarian in treating any boar suspected of transmitting the problem.

Is the return erratic throughout mid-pregnancy? If this is the more typical pattern it could still arise from the same problem as described above. This irregular sequence might also arise from trauma – in particular from stress caused by bad floors, environment or bullying.

Certain epidemic diseases of a transitory nature, such as swine influenza and TGE, can contribute to an erratic infertility problem causing both irregular returns to service and inhibiting onset of oestrus.

REPRODUCTIVE PROBLEMS, OR S.M.E.D.I.

Although referred to at various points in the text to date, separate reference to these organisms has been delayed until the husbandry factors have been fully covered. This is proper because reproductive viruses are used as a convenient excuse for reproductive failure without a thorough check of all other management procedures being made.

Organisms which disrupt reproduction are thought to be numerous. Add to this the fact that they will produce different effects according to the stage of the reproductive cycle when the sows or gilts are challenged and the picture is confused still further. This is made even more difficult because the differing ages of sow in the herd means that almost every animal will have a different antibody or immunity status towards the disease.

Such differing effects have given rise to the common pigman's mnemonic for conditions caused by these organisms—S.M.E.D.I. Analysis of this allows a full understanding of the possible effects of parvo-virus infection of a herd and one or a combination of these effects may be seen simultaneously in a unit:

S.   —Stillbirths
M.   —Mummification
ED   —Embryonic death
I.   —Infertility.

Vaccination is now possible, normally using a dead proprietary vaccine. Vaccination is most effective where gilts are vaccinated as near to first service as is practicable and then again within seven days of farrowing. In a herd where there is a major disruption to reproductive activity sows might also be vaccinated within seven days of farrowing.

The weapons at the British pigman's disposal are few and crude, but their diligent use may help maintain a high level of immunity.

In addition to vaccination it may also be useful to attempt to influence a higher level of immunity by the following husbandry techniques:

The main aim must be:

● To challenge non-pregnant gilts and sows with any debris from the farrowing house. This includes sows suckling their litters which have farrowed more than four days previously. It may be helpful to macerate afterbirth etc. from sows suspected of having an active dose of the organism and feeding this to non-pregnant sows and gilts.
● To give close contact between sows thought to be infected and maiden gilts.
● To *avoid* challenging served sows or gilts because it may cause an unpredictable response and a rise in temperature with subsequent reabsorption or abortion.

There is no doubt that scientific knowledge about reproductive viruses is incomplete and time will yield greater information to assist in overcoming the effects of this group of organisms.

In addition to those organisms thought to effect the reproductive organs, other viral bodies which invade the respiratory or digestive organs may well indirectly impair reproduction by causing general debility in the breeding animal. Listed amongst such organisms are P.R.R.S., Aujeszky's disease, classical and African swine fever, coronavirus, swine influenza and TGE.

Remember
It is now widely accepted that these groups of viruses grouped under the S.M.E.D.I. acronym are unlikely to have a major influence upon onset of oestrus as returns to service which occur in the month following first mating. Nor is abortion a likely manifestation of infection by reproductive viruses although very premature farrowing may be associated with it. Further, unless the herd also displays a constant pattern of mummified piglets at birth as well as an erratic fertility pattern then the S.M.E.D.I. group is unlikely to be implicated and the husbandry factors described are more probable causes.

Immunity can only come from vaccination or by acquiring immunity from the mother or by coming into contact with infected animals or debris.

Different strains of viruses are likely to be involved so naturally acquired immunity might be shortlived and the 'dose' of immunity, or its effect, will vary according to the overall status of the herd at the time, the antibody status of the gilts' dam and the age of the gilt when purchased or transferred to the breeding herd and the number of non-immune females introduced into the herd.

However, just because it is difficult to break the infective cycle of reproductive viruses (even by hysterectomy), this should *not* deter the pigman from closely following the recommendations given for pig acclimatisation.

Chapter 12

The size of litters born is one of the major topics of conversation whenever pigmen meet. It is taken as a kind of status symbol and there is a justification for this in that the number of pigs born does reflect, to a large extent, the pigman's input in terms of attention to detail and diligence.

Frequently the extent of litter size quoted is misleading. We can imagine that litter size has improved whereas we are really seeing a response to an increase in the average age of a herd. Thus it is first necessary that the pigman understands that failure to regulate the average age of herd will give rise to a varying potential for litter size, (see Chapter 11).

Next, a problem with numbers of pigs born may not be a fecundity problem. Where records do not include an indication of stillbirths (and mummifications), it is possible to conclude that litter size is small, whereas it might well be that the problem is not total numbers, but the proportion of stillborn pigs, and this will be discussed later.

Furthermore, before attempting to 'hunt down' the causes of low numbers born, it is desirable first to determine whether the problem affects the whole herd or individual sows. This demands, again, the accurate maintenance of records and careful interpretation. First, the range of litter sizes should be examined with particular emphasis on the proportion of sows which produce less than seven pigs in total and that portion of the herd which exceeds thirteen pigs in total. The information will point the way to the kind of actions required. A herd should have an increasing pattern of litter size up to around the fifth litter (see Chapter 11). If a study reveals the kind of pattern shown in fig. 12 then it can be assumed that:

● Gilt management is at fault and the feeding regime, in particular, is probably giving rise to too great a bodyweight change during first lactation, so that young sows are served in poorer condition after the first weaning.
● Boar management is poor.
● Service control is inaccurate, so that sows fail to reach their optimum.

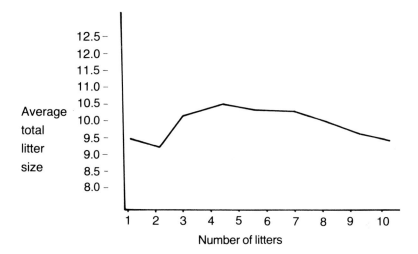

Fig. 12. Undesirable pattern of litter size

● Recording of litter size is not precise.
● There is a failure to compensate for environmental shortcomings by good husbandry and adequate nutrition.

Thus, a change in litter size in a herd with a farrowing pattern as shown in the diagram, will be aggravated by a change in herd age when there is a large proportion of first- and second-litter farrowings.

Such results indicate a total review of gilt management from selection or delivery through to weaning, and the routines recommended in Chapters 10 and 11 should be reconsidered.

If a parity analysis does not reveal such a problem of particular groups of sows or herd age influences, the extent of the individual sow problem may then be considered. There is a considerable discrepancy between the potential litter size and that normally achieved in any case and most of it, as stressed, is influenced by the pigman.

GENETICS AND LITTER SIZE

The potential for litter size is set by genetics, but this is largely outside the control of the pigman. The owner, or whoever makes the decisions, will choose the type and source of replacement stock and the pigman's task is to attempt to achieve the best results from that animal.

However, the geneticist may well produce animals with a greater capability to produce large litters in the future from careful selection within breeds or from the introduction of new breeds. In the meantime, the pigman must ensure that he:

● optimises the hybrid vigour benefit of the first-cross sow over a second or multi-cross sow by planning herd replacements so that there are always sufficient first-cross gilts to hand;
● carefully records and identifies stock to guarantee that there is no confusion between the first-cross and slaughter generation gilt.

It is possible to increase litter size slightly by serving the first-cross sow by a boar of a breed which is different from those used in producing the first-cross sow. The pigman must be aware that the owner or manager will be reluctant to use a third breed because it will almost certainly be non-white skinned, and some outlets do not favour 'blue' or tinted pigs. Almost any other breed grows less rapidly than the Large White and Landrace types, and it may well increase backfat levels compared to the two major white breeds.

Thus, the pigman should understand an owner's indifference to the possible litter size and piglet vigour effects of a boar line unrelated to the typical British Large White × Landrace sow.

The reasons for the superiority of some breeds as far as litter size is concerned are not clear, although some scientists believe that the size of the womb may be influenced by the body shape of the pig. In particular that the longer breeds or strains may have a greater length of 'horn' of the uterus and surface area to allow the attachment and development of more pigs. This may lead, in the future, to separate development of female lines and sire lines to produce specific carcase shapes. This will require even more attention from the pigman to ensure that the offspring from these, probably even less prolific, sire lines, do not get retained for breeding and depress litter size potential.

What Litter Size Should Be Achieved?

A normal, healthy hybrid sow probably sheds about thirty eggs and a gilt perhaps twenty-four. A well-managed sow and gilt served according to the recommendations given should achieve a 90–95 per cent fertilisation rate. So why is it that twenty-plus pigs are not born? Breed effect has been mentioned, but this is probably an

example of nature's 'safety net' provision to ensure the maintenance of a species under the most adverse conditions.

Thus, the pigman should set out to decrease the adverse conditions.
It should be expected that every sow in the herd is capable of
producing at least 90 per cent as many pigs as the most prolific sow in
herd. While this might be considered to be the aim, more
realistically the aim for a herd with a well-regulated spread of sow
age will be *twelve live pigs and no more than 5 per cent stillbirths per
litter*.

A *total* of less than eleven pigs born must be considered
unacceptable; in this case a comprehensive review of management
procedures and herd health is needed.

WHY DO ONLY HALF THE NUMBER OF FERTILISED EGGS SURVIVE?

Aside from genetics and the age of sows the number of eggs produced,
fertilised and retained may be affected by the husbandry features
mentioned in the points below.

How long since the sow has farrowed?

The shorter the interval between farrowing and remating, the less
time there is for the womb to return to its normal state. Although
there is some variation from herd to herd it appears that providing
the animal is in fit bodily condition and her environment is
satisfactory:

● there should be little effect upon subsequent litter size providing
   that the interval between farrowing and service is no less than
   eighteen days.

The difference between herds in this respect is probably due to
the variation which exists in the conditions on farms and the failure
to compensate for the needs of the individual sow by more generous
feeding regimes in particular.

It is important to check closely that an erratic litter size pattern is
not the result of a proportion of sows being weaned much earlier
than the majority of the herd.

Is the sow weaned in satisfactory condition?

It is not sufficient for the pigman to consider the appearance of a
sow at weaning in relation to other sows. If a sow is fitter than
average at point of farrowing, she may still appear satisfactory at
weaning, but will have undergone a greater weight loss when

suckling. In other words, the *degree of change* can have an influence on her recovery and subsequent performance, which compares to the more obvious appearance of a thin sow.

Thus, the need to wean sows with a satisfactory level of backfat so that they commence the next pregnancy in suitable condition is vital.

This calls for a generous level of feeding to sows during lactation. While the programme for feeding at farrowing time is discussed in Chapter 13, once a sow has been farrowed for four or five days she *must* be fed generously.

The modern pigman should *not* be too constrained by rigid feed scales. The amount required to avoid excessive backfat loss in a sow will vary widely, so the feeding regime should be:

That amount of feed it takes to minimise a loss of condition.

There is no merit in low sow feed usage if it inhibits full sow output. The pigman should remember that:

● there is little chance to make good lost backfat between weaning and service;
● once a sow has started to lose backfat it may take several weeks to arrest that trend regardless of the amount of feed that she is subsequently offered.

What can be done if the sow will not eat sufficient feed when suckling?
The inability to get sows to eat sufficient feed to prevent excessive backfat losses in lactation is common. It is so vital that this be overcome that the sequence shown on page 129 should be carefully considered for healthy sows.

Judgements concerning sow condition and weight change are stressed again in Chapter 13.

What can be done with sows which eat too little after weaning?
Although relatively little can be done to recover lost condition in the relatively short period between weaning and service, it is still desirable that the pigman encourages the sow to eat large quantities of feed in order that:

● the drying-up of the udder be speeded-up;
● sufficient energy intake occurs to encourage high ovulation rate.

The pigman should therefore check:

● water uptake;

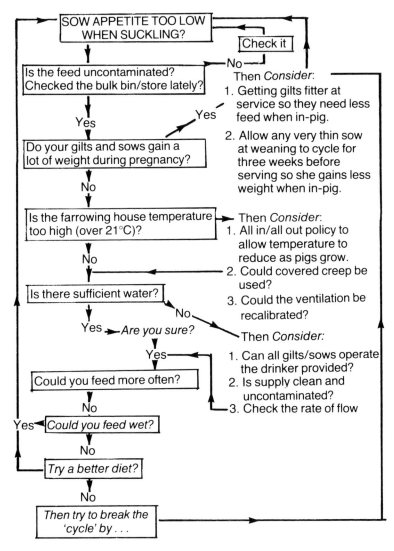

• that bullying is not taking place—there is some evidence that sows penned individually have bigger litters even though they may take longer to come 'on heat'.

Sows may be encouraged to eat more by:

• feeding more frequently. Up to the point when the sow has been

served for the last time, you might consider offering a second (or third), feed following the morning and afternoon service checks.
● offering feed as a wet mash if trough design permits.

Is the house temperature satisfactory?

Although feed intake can be adjusted to help offset low temperatures to some extent it is important to attempt to reduce the effects of high temperature —prolonged periods in excess of 27°C not only reduce ovulation, but increase the risk of embryo loss after service. The pigman should be ready to:

● adjust vents to increase air movement;
● deflect incoming, faster-moving air over the pigs;
● dampen floors to increase evaporation;
● provide shelter if outdoor runs are provided.

As mentioned in Chapter 9, prolonged periods of high temperature also reduce boar fertility which may have a further adverse influence upon litter size.

Is the sow stressed?

The importance of separating timid sows from batches has been referred to in Chapter 11. You should also check whether floors and slats are damaged in individual sow pens, because a sow discomforted is less likely to conceive a large litter. If a sow spends much of the day sitting rather than standing in the stall, she may well be giving a signal that she is uncomfortable and this should be noted and attention given to the flooring. Leaking troughs and drinkers are another cause of harmful discomfort that should be checked and eliminated.

Are the boars overworked?

The need to regulate boar use has been previously stressed and may be referred to in Chapter 9—also see notes below on service routine.

Sow health normal?

Any condition which causes a rise in the animal's body temperature can result in embryonic loss, if not abortion. Therefore, any of the following factors which can give rise to body temperature should be considered and, where necessary, brought to the attention of the veterinarian:

- mouldy or contaminated feed;
- erysipelas;
- cystitis or nephritis of kidneys;
- pneumonia and other illnesses;
- joint infections.

Is the service routine adequate?

The importance of a carefully monitored service and sow checking routine has been stressed. Poor control will give poor results and this area may contribute more than many others to poor litter size.
It is particularly important to ensure:

1. That a proper physical mating occurs.
2. That the sow is served close to the time when it is likely that the eggs are being released—which is approximately twenty-four hours after 'heat' has commenced. The accuracy of this timing will be increased by:

- twice-a-day checking by the boar;
- serving at least twice and, preferably, three times during oestrus to offset the inevitable doubt when oestrus *actually* starts;
- if, for reasons of boar availability, only two services per sow can be made and AI is not available, then the service to miss would be the first one after oestrus is observed as eggs are shed some 24 hours following the time that the female normally accepts the male (see Figure 11).

Providing a herd has no history of venereally spread disease, a potential advantage from the use of different boars when repeating the service is that it may increase litter size.
The reason for this increase in litter size may arise from a combination of:

- increased 'motility' of 'mixed' sperms;
- better resting where boars do not have to serve two to three times over a 24–36-hour span.

However, the pigman should be aware that he will only achieve the benefits from serving two to three times and cross-mating using different boars where *he has sufficient number of boars to avoid over-use*.
It is probably that those herds which report no benefit from multiple services have failed to increase the number of boars in order to compensate for the higher rate of usage when frequency of sow service is increased.

Remember
The boar:sow ratio has its limitations. In a 200-sow herd with a ratio of 1:20 sows, if three-week weaning is practised there may be a need for thirty-three individual matings a week. This exceeds most recommended rates of boar use (three to four services per boar per week).

WHAT CAUSES STILLBIRTHS?

One of the prime causes of stillbirths is failure to observe farrowings. This is another way of saying that some pigs which are recorded as stillbirths may have been saved if closer supervision had been given.

Stillbirths are of four main types and it is desirable to generate accurate records to determine which category is causing the main problem. These four types may be:

1. 'Mummified' pigs—these appear as black–brown partly-decomposed foetuses and would have been the result of some 'interference' during the second or third month of pregnancy.

2. Death before farrowing—these may be perfectly formed piglets except that they take on a 'sunken-eyed' appearance. These deaths probably take place in the last week or so of pregnancy.

3. Death during farrowing—these are also perfectly formed pigs and a prime cause of this category is the extension of the farrowing period causing the later born piglets to be suffocated prior to birth.

4. Suffocation in the membrane—about 5 per cent of pigs tend to be born wrapped in a 'sac' of cleansing from which few have the initial strength to separate themselves.

There is a tolerable expectation of stillbirths in all mammals and particularly so in multiple-birth creatures like the pig. However, the pigman should consider the stillbirth rate too high where *total stillbirths exceed 4 per cent of the total numbers born* (i.e. should be *less* than half a pig per litter on average).

What can be done about mummification?
Mummification results from some agent causing the rejection of one or a number of embryos. The most likely cause of mummification is a challenge from reproductive viruses which may cause a progressive deterioration giving a range of sizes in the dead piglets presented (see Chapter 11). The main way in which the pigman may help to control this problem is to attempt to raise the immunity of pigs

within the herd by the use of the measures previously outlined.
The pigman should also suspect:

● Dietary influences—failure to match environment and sow condition problems with sufficient dietary energy and the feeding of contaminated feed can increase the problem.
● Physical damage—fighting or other accidents may also predispose the presentation of mummified pigs.
● Erysipelas or any other condition causing a rise in body temperature.

What can be done about piglets which die before farrowing time?

These may be caused by a variety of influences and, although reproductive viruses cannot be ruled out, the likely causes may include:

● Physical damage when moving the sow into farrowing quarters.
● Excessive excitement caused by feeding very low levels just prior to farrowing or an ultra-excitable temperament in the particular animal.
● Energy deprivation—increase energy levels, perhaps by special supplementation, in the final week of pregnancy.
● Age of sow—older sows have a tendency to produce a greater proportion of still-births.
● Design of farrowing pen—where sows are penned with their rear to the feed passage some may attempt to turn around when first penned in the house.
● Insufficient handling of the gilt—gilts separated for the first time from their pen-mates should be closely monitored by the pigman and sedated if excessively distressed.

Clearly the pigman should be ready to act in handling and soothing the excitable sow or gilt and should take care to avoid excessive feed restriction in this period.

What can be done about piglets which die during farrowing?

Many of those piglets lost in this category must be considered as 'savable'. The first probable contributor is the time taken to farrow. The pigman can help to reduce this problem by:

● Trying to ensure that the sow or gilt is physically fit at farrowing.
A lame sow which has been inactive due to some damage or injury must be considered a likely 'slow farrower'.

- Ensuring that the sow is not constipated. Constipation also has the disadvantage of raising the sow's temperature which should be avoided at this time, as it predisposes the sow to inappetance and possibly, farrowing fewer.
- Overfeeding prior to farrowing. A distended gut can constrict the reproductive tract, slowing the passage of the piglets.
- Avoid excessive weight gains in pregnancy which can cause a 'sluggishness' at farrowing time.
- Stress caused by a sow being uncomfortable in the farrowing crate. This again generates a need to ensure that all surfaces in the pen are non-slip and non-abrasive. Further the sow should be housed well in advance.
- Also under the heading of stress comes a consideration for the use of bedding. Behavioural studies suggest that some sows may have prolonged farrowings because of being deprived of the opportunity to make a nest by the withdrawal of bedding.

The pigman can further reduce the number of losses due to a protracted farrowing by close supervision of the sow and being on hand to ensure that birth occurs over a reasonable period of time, which should not exceed 2½ hours with regular intervals between piglets (see below).

The pigman's presence should, however, not distract the sow and this is only prevented by regular handling of the sow or gilt during the period she is housed in the farrowing crate. Although no pigman has time to 'spare', he should take every opportunity to 'settle' gilts and sows prior to farrowing by rubbing and speaking to them. If the need then arises to assist at farrowing the animal will be less alarmed.

If the pigman is on hand he may take the decision to assist with farrowing (see Chapter 14), where there are obvious signs of distress in the sow and/or a long interval (over twenty minutes) has occurred with no births and the sow is attempting to pass pigs. To assist in deciding whether assistance should be given, those attending farrowing should note on the recording card or chalk onto a suitable surface the times when the pig last was presented or treatment was given. Drugs may also be used, under veterinary supervision, to speed farrowing to prevent losses due to asphyxiation.

Supervision of farrowing may be made easier to organise where farrowings can be predicted and the use of drugs to synchronise farrowings may help in this respect. These are discussed in Chapter 14.

Remember
Where litter size appears to be low, the pigman must satisfy himself that he is not interpreting the records falsely and must then re-check his management routines.

Chapter 13

Methods of sow feeding seem to have their roots in folklore. On few other matters with which the pigman has to juggle have there been so many opinions expressed over the years. Perhaps the pigman's willingness to try different regimes stems from his genuine anxiety about sow performance and sow condition.

It is generally unhelpful to consider sow feeding in terms of so many pounds or kilograms per day. It is far better to consider sow condition and to adopt a philosophy which *gives an economic output for every bit of feed fed rather than restricting feed intake and hoping for an economic return.*

In practice, changes in sow feeding are too often made as a response to the animals' thinness. If more than 10 per cent of sows at weaning are too thin then a more strategic look at sow feeding is called for. In addition, a relatively large change in sow condition, particularly during the suckling period, even if the sow appears to be in satisfactory condition at weaning, can have an adverse affect on her performance similar to that of a very thin sow.

How Can Sow Condition Be Judged?

Success in treating sows as individuals has been observed on many units. This does not automatically mean that they have to be individually housed but rather an assessment is made of their condition and feed scales are adjusted accordingly.

The pigman should, probably in conjunction with whoever makes decisions on the unit, undertake a monthly check of the dry sow herd to:

- establish the proportion of sows in unsatisfactory condition;
- particularly check the condition of newly weaned sows;
- review feed levels being offered.

Where sow condition is a problem a more formalised system of sow condition scoring may be used. This technique has been developed because the extent of the problem in the national herd has been acknowledged. Applying a condition score to each sow does focus the pigman's attention on this subject and, by doing so, increases the likelihood of a response to the problem being made.

Obviously, sows in different stages of the reproductive cycle will be in slightly different condition—one that is about to farrow will naturally appear in better body state than a sow from which a large litter was weaned this morning! However, the pigman should set out with the aim of ensuring that few sows are weaned showing the following features:

- lack of good cover over the pin bones each side of the tail;
- they should not have a hollow loin or prominent pelvic bones over the loin;
- the spine and ribs should not be visible and should only be felt by firm hand pressure.

A more detailed condition-scoring chart is shown on page 138 and may be used to give greater refinement than the generalised comments given above. If the chart below is adopted for use it is recommended that *no more than 10 per cent of sows at weaning should fall in categories 2 or less*.

In addition, if the chart is used for the whole herd, the allowance for the varying stages should apply; for instance, sows in later pregnancy should be judged more harshly than those most recently weaned. Further, it is uneconomic as well as undesirable to categorise too many sows in category 5. Once sow herd condition has been returned to a satisfactory status, the more formal scoring of condition may then give way to the regular monthly routines suggested.

The rate of change in the condition of sows of a satisfactory general appearance can only be detected by weighing sows and recording their weights at consistent times.

Weighing does necessitate the movement of sows and the correct equipment. It is usual only to use this method of logging relative change when sows are moved into the farrowing quarters and being weaned, due to the inconvenience of moving pigs. Condition scoring can be conducted in the sow's own pen and needs no equipment. However, weighing can be a useful aid to full productivity in those herds which appear to have few thin sows.

If sow-weighing is practised, the pigman should:

- carefully record the information on each sow's card;
- expect that there will be a cumulative weight gain up to around fourth or fifth farrowing;
- expect that there will be an inevitable weight loss between farrowing and weaning, but the aim should be to keep this change to a maximum of 40 kg and, preferably, less.

Category description	Tail head appearance	Loin, ribs and flank appearance	Backbone appearance
0. Emaciated No sow should fall into this group.	Deep cavity either side of tail with pin bones very prominent.	Individual ribs and pelvis bones very obvious. Flank hollow.	Individual vertebrae obvious.
1. Poor Few sows should come within this group.	Cavity around tail can be seen, slight cover over pin bones.	Ribs and pelvis just covered, but flanks pinched and loin narrow.	Individual vertebrae prominent.
2. Moderate 10% of newly weaned sows may be categorised as such.	Pin bones covered and little sign of cavity around tail.	Ribs and pelvis covered, but may be felt when sow is handled.	Some cover over spine.
3. Good The category to aim at for weaned sows.	No cavity around tail and pressure is needed to locate pin bones.	No hollowness in flank, and ribs well covered.	Vertebrae will only be felt with firm pressure.
4. Fat The category to aim at for sows soon to farrow.	Evidence of some fat cover around the tail head.	Ribs and pelvis cannot be felt.	Backbone cannot be felt even with hard pressure.
5. Very fat Unnecessary.	No greater fat cover can be achieved at any point.		

(After ADAS Scheme devised by J. Deering).

Because sow weighing is most useful when used to measure relative weight change, it can only be said to be really helpful when a lifetime log of the individual sow's weights are known. For example, a gilt weighing 110 kg when first served will follow a different pattern of weight change, and probably ultimate mature body size, to a gilt first served when weighing 130 kg.

The table on page 139 may be used as a guide in those herds wishing to consider use of the technique.

Some producers are trying to use ultrasonic backfat measurements to assist in interpreting the change in sow condition, but this is difficult to interpret and, like sow weighing, is only really of use if checked over the sow's entire lifespan.

Status of gilt or sow	Approx. weight (kg)	Approx. weight (kg)	Approx. weight (kg)	Approx. weight (kg)
First service	105	110	115	120
First farrowing	150	155	160	165
First weaning	115	120	125	130
Second farrowing	165	170	175	180
Second weaning	125	130	135	140
Third farrowing	180	185	190	195
Third weaning	140	145	150	155
Fourth farrowing	195	200	205	210
Fourth weaning	155	160	165	170
Subsequent farrowings	210	215	220	225
Subsequent weanings	170	175	180	185

Remember

It is difficult to be rigid in interpreting condition scoring and weight changes, and the simplest approach is to monitor sow condition continually and attempt to keep to a minimum the proportion of sows in less than satisfactory condition.

WHAT CAUSES SOWS TO BECOME THIN?

There are three main causes of thin sows:

● too little nutrient intake, particularly energy, for the particular circumstances on the unit;
● faulty distribution of feed over the gilt or sow lifetime;
● some health condition interfering with the sow's response to her diet.

Energy intake may be too low where:

● housing is too cold—see Chapter 7 for checklist;
● competition is too great—individual feeding or separation of the individual should be considered;
● diet is low quality;
● weaning age is extended;
● sows are too thin due to the other two categories below.

Distribution of feed over the sow or gilt lifetime may be wrong where the following circumstances apply:

1. Gilts are fed so that they gain excessive weight during pregnancy, often as a result of inadequate backfat reserves at service, see Chapter 10.

2. Their appetite or intake is too low during suckling, resulting in a steep decline in condition, see Chapter 12.

3. The combination of poor appetite and wide changes in sow condition leads to a need for high feed levels, to correct lactation weight losses, in following pregnancies. If occasional sows or gilts are too thin at weaning, the pigman should consider allowing them to regain condition before re-mating, rather than attempting to replace backfat in pregnancy. Providing this does not interfere with the service and farrowing programme, it can be used as an 'occasional' technique in attempting to correct excessive fat losses prior to service and to break the cyclical loss:gain pattern which probably started with the gilt.

4. Some modern high-energy diets may be slightly less palatable particularly when badly stored, and intake may be depressed as a result offsetting the advantage of these 'denser' diets.

Other influences on sow condition may include:

1. *Health*—sows which have pneumonic conditions, are slightly lame, or have undergone farrowing fever checks, may have a depressed appetite and undergo excessive weight loss.

2. *Parasites*—excessive worm burdens may interfere with feed utilisation and a regular programme of worm treatment must be planned with the veterinarian and rigidly adhered to.

Remember
The strategy for sow feeding must be:

● serve gilts in good condition;
● do not overfeed in pregnancy;
● feed generously from a few days post farrowing to service;
● if any sow is too thin, try to put it right prior to service or anticipate that she will become thin again during her next lactation.

The Tactics for Sow Feeding

To apply these general principles the pigman must be ready to adjust feed scales to deal with those sows who still do not respond to the procedure outlined.

Where sows get towards the *end of pregnancy* in only moderate body condition they are likely to:

● have smaller litter birthweights and lower vigour which reduce survival expectations and probably weaning weights;
● have lower milk yields.

To overcome these specific problems, consider feeding more energy in the last month of pregnancy. This can be achieved by:

1. Increasing feed scale by one-third.
2. Using a supplementary energy-booster supplement in addition to the normal feed scale which, being high-oil, increases intake without excessive change in the bulk consumption.

It should be stressed that changes in piglet birthweight may be difficult to measure, but where sow condition is less than satisfactory, survival rate may improve. This technique is most successful with the thinner sow and where the standard diet used is not, itself, already particularly high in energy.

Concerning the period *immediately prior* to and for the first *few days after farrowing* much contradictory advice has been given. The facts are that high levels of feeding at this time can cause the gut to become filled and make farrowing more sluggish, and most sows naturally produce sufficient milk initially without extra stimulation from feed levels. Indeed, higher feed scales may be implicated with mastitis at this stage.

The problem is that over-restriction at this time:

● can lead to excessive sow excitement and still-births due to physical loss.
● can lead to reduction in vigour of piglets at birth and increased mortality.

The advised tactics for the period between housing in the farrowing quarters and, say, the fourth day after farrowing are: *to feed no less than the normal amount given during pregnancy and to make any ration change with care.*

Feed levels between *four days after farrowing and weaning* must be geared to maintaining good sow condition and rapid piglet growth. It may be helpful to set a minimum feed scale and to adjust upwards depending upon the individual sow and her litter. At this stage the setting of a minimum, rather than a maximum, will help to draw everyone's attention to the need to wean a fit sow (see Chapter 11 for influences upon appetite when suckling). It is unnecessarily complicated to adopt 'so much for the sow, plus this much for each pig', because every sow may react differently to a pre-set feed scale in any case. Consider adding wet bran mash or pig oil (140 ml per

sow per day) to the sow's diet prior to farrowing to avoid constipation. Dung condition must be regularly examined, particularly
where lower fibre sow diets are used and where dry sows are housed
in bedded accommodation and are moved to unbedded farrowing
pens.

Any sow refusing feed after farrowing should be given immediate
attention. As well as checking the sow's temperature, water supply
and udder condition it is valuable to release her from the pen and
allow her to roam, preferably on grass. This exercise usually
encourages her to dung and urinate and often triggers a spontaneous
improvement.

At around *weaning time* there is no justification for a large
reduction in feed levels. Some producers consider that this reduces
the risk of mastitis, but it is rarely true for reabsorption and drying
up of the udder are speeded up where feed levels are high and the
litter removed. Indeed, so short is the re-breeding interval that
every effort should be made to maintain high feed levels to ensure
that sows are served in the most receptive body condition.

Where, despite all efforts, a sow is *still thin during pregnancy*,
every effort must be made to correct the deficit—even if it *does*
mean that she may well lose too much backfat again next time
around.

Other Sow Problems

In briefly debating the common sow conditions below, the aim is not
to turn the pigman into an amateur veterinarian, but to create
awareness of how pigmanship creates conditions which makes
incidence of the conditions more likely to occur.

Mastitis

This can be a killer disease. It can be caused by a variety of bacteria,
but almost always has its origin in hygiene. Coliform mastitis may
affect the whole of the udder or large areas whereas some other
bacteria affect single teats. Occasional cases must be anticipated
but, because they cause a rapid rise in temperature and loss of
appetite at the very least, the pigman should quickly refer cases to
his veterinarian. The udder becomes hard and inflamed, so piglet
losses are more likely where it occurs. If more than 5 per cent of
sows or gilts which farrow have mastitis, a major investigation is
called for.

Mastitis is less likely to occur where:

● the sow does not lie in damp pens caused by poor drainage;

- the sow does not lie in damp soiled bedding;
- the sow does not lie in damp pens caused by leaking trough and drinkers or pens which have not dried after pressure washing;
- the sow does not lie in pens with badly pitted or broken floor surfaces;
- the sow does not lie in pens which are badly cleaned out;
- the sow is not expected to drink water from a badly managed water supply;
- the sow is not overfed around farrowing time;
- the teats are not damaged by a badly designed floor or slats;
- piglets' teeth are clipped at birth, avoiding udder damage.
- flies are well controlled so reducing the risk of transferring bacteria.

Agalactia

This is where the sow has little, or no milk. It is thought to be complex in its cause, but is associated with some kind of shock which reduces the stimulus for the sow to release her milk.

It may be less of a problem where:

- the sow is fit at farrowing and undergoes no major shock—so sow condition and handling prior to farrowing may help;
- farrowings are observed and not allowed to become abnormally extended;
- pens are clean and dry so that the sow is comfortable at farrowing.

Poor Milking

Occasionally sows' udders dry up prematurely, causing poor pigs within a litter and increasing the need to foster piglets to other sows. This can occur even when no signs of mastitis are present. The pigman should ensure that signs that piglets are not growing well do not result from chilling/scouring because such piglets will not suckle well and will thus predispose the udder to drying up.

Poor milking or premature drying up of the udder is less likely to occur where:

- the sow is normally and vigorously suckled, particularly in the first week after farrowing. Thus, care should be taken when fostering very young pigs to ensure that, if all small ones are placed on one sow, there is sufficient stimulus to encourage normal udder activity.
- the sow is not overfed immediately after farrowing as over-production of milk may cause premature drying.

- the sow has adequate water — some may have difficulty in using a particular type of drinker. The pigman must be aware of this and be prepared to give more water by hand.
- the crate dimensions are not restricting access to the teats by piglets causing reabsorption and premature drying of the udders.
- the sow's past records have been used to ensure that excessive age or a persistently poor milking sow are not the reasons for this problem.

Metritis/Vaginitis

These arise from infections of the womb or vagina by bacteria. Although quite commonly associated with farrowing time, they may also arise during pregnancy. A discharge is seen and the rise in temperature has similar effects to mastitis. In addition, it is thought that affected sows may undergo temporary or permanent infertility. Metritis is more common as a result of damage at farrowing time and may be predisposed following assisted farrowing, but should never occur in more than 2 per cent of the herd.

Metritis/Vaginitis are less likely to occur where:

- great care and hygiene are practised with assisted farrowings;
- the same cleanliness and hygiene routines recommended under MASTITIS are followed;
- dry sow pens are cleaned down to reduce the bacterial burden and risk of ascending infection—particularly stall and tether standings. Thus slatted pens should be scraped clean regularly to prevent accumulation of muck, and both they and solid standings should be pressure-washed occasionally.

Urinary Infections

Bacterial infection of the kidneys (nephritis) and bladder (cystitis) are quite frequent, but should not exceed 2 per cent of the sow herd. They are normally noticed during pregnancy when the sow loses appetite and pus or blood appears in the urine. Infected sows rarely retain pregnancy or fully recover.

These infections are less likely to occur where:

- pen hygiene is good;
- control of services and good monitoring allow boar checks to be made to minimise chance of spread at service time.

Difficult Farrowing

Despite sows being thoroughly prepared for farrowing, some do suffer uterine inertia which is a failure of the normal farrowing process.

This is unlikely to be transmissible and is less likely to occur if:

- the sows are young;
- the animals are well settled in their farrowing quarters and are used to the pigman's presence;
- piglets are not too large (which can occur in small litters);
- sows are not overfed prior to farrowing;
- the sow is comfortable because the pen is in good repair and especially because bedding is available to encourage nest building;
- records are consulted, so that those experiencing this condition may be carefully monitored at farrowing time.

Savaging

The reasons for sows savaging their piglets at birth are not clearly understood. Sows and gilts are less likely to savage if:

- they are regularly handled prior to farrowing, particularly in the case of gilts;
- records are used to predict likely cases;
- sows are not excessively fit at farrowing;
- the serving of gilts very young in age is avoided;
- gilts or sows are sedated if an inclination to savaging is observed. Injections of sedatives can be used and some pigmen find two litres of beer prior to farrowing a good aid to settling the excitable gilt.

If the sow is savage, the pigs should be carefully shielded from her and only allowed to suckle under supervision. Complete withdrawal of the litter and sedation may extend the farrowing period, so close supervision until after farrowing and the sow has settled down is necessary. Muzzling of the sow or gilt may be practised.

Vulva Haematoma

This follows damage to the vulva and the appearance of a 'blister' filled with blood and fluids.

This is less likely to occur if:

- piglets are not too large at birth;
- a sow is farrowing as opposed to a gilt;
- great care is taken if farrowing assistance has to be given;
- it is remembered that a very excited gilt may crush her vulva against the rear gate of a crate or stall, so the pigman is ready to

adjust gate settings and even hang a straw-filled bag on the rear gate of a pen if a gilt is very excited when first penned alone.

Prolapse

This problem may occur in spasms in certain herds and the rectum, uterus or cervix might be involved.
It is less likely to occur if:

● rear parts of individual crate and standings are not slippery;
● there is no steep step in the length of the standing;
● the sow is not 'straining' excessively at farrowing time;
● the sow is not constipated;
● correct levels of calcium are incorporated in the feed and the ratio between calcium and phosphorous is in proper balance (refer to feed advisor/supplier is prolapse uterus occurs more than twice per 100 farrowings);
● the slope of the pen floor is not too excessive as to cause extreme pressure upon the cervix.

Worms

A range of internal parasites can, if unchecked, have a marked effect upon sow condition and performance. It is necessary to consult with a veterinarian on the type of worm treatment used because products vary in their method of control and in the range of parasites affected.
Worms are less likely to be a problem where:

● dung samples are regularly, i.e. quarterly, checked for the presence of worms and coccidia;
● an appropriate product is selected and its use regularly monitored to ensure that no animals are 'missed' (in some herds it is sufficient to worm sows prior to farrowing only in order to achieve good results, in others it may be necessary to worm growing pigs in addition;
● all incoming stock is wormed prior to being brought into the main herd from the acclimatisation/quarantine premises.

Mange

This skin disease is spread by a small mite living on the pig. Ears and other areas where the skin is soft are favourite areas of infestation and the animal becomes irritated and restless with patches of rubbed, encrusted skin.
Mange may be less of a problem where:

- skin is scrubbed, removing crusts from obvious areas with warm water and antiseptic solution;
- treatment is applied in an approved method (injection, pour-on or spray-on), in accordance with the manufacturer's recommendation;
- repeat/follow-up applications are made at the correct interval in order that the 10-15 day life cycle of the mite is broken (survival is around three weeks away from the host). This particularly refers to the less residual spray-on treatments;
- it is ensured that all adults are treated to reduce risk of rein- fection;
- all weaners are dipped at weaning time;
- periodical pressure-washing and disinfecting of the dry-sow standing, boar pens, service areas etc. is carried out.

Remember
The pigman should not need to diagnose or know the correct treatment for these conditions, but he should be able to keep occurrences to a minimum.

SECTION VI

Chapter 14

Pre-weaning Management
Including checklist for:
Piglet survival
Prevention of overlying
Prevention of starvation
Fostering techniques
Prevention of chilling
Synchronisation of farrowing and supervision of farrowing
Assisting farrowing
Revival of new-born piglets
Artificial rearing
Navel bleeding
Splayed legs
Routine tasks
Joint ill, blind anus, knee damage, scours, rhinitis
Weaning time management
Creep feeding.

Chapter 14

Previous chapters have dwelled upon the routines and activities needed to achieve large litters at a regular and predictable interval. It is proper now to consider piglet survival because, having achieved a satisfactory mean litter size of around twelve live pigs, attention must be focused upon ensuring that the maximum number reach sale weight. This is, again, an area where even the youngest member of a team can significantly influence performance and where close supervision and accurate recording will yield large benefits.

Losses of piglets between birth and weaning have shown an obstinate refusal to improve since national statistics have been available. Relatively few herds manage consistently to achieve single-figure percentage results. Yet, some herds show levels of pre-weaning mortality around 5 per cent of all livebirths, so it can be presumed that this must be an achievable figure and one to which all pigmen must aspire.

The disappointment in being unable to influence more than a handful of sows to turn more than half their fertilised ova into live piglets means that a greater effort should be made to assist the survival of those pigs we do get born.

The greatest influences upon piglet mortality are overlying and starvation—which may account for up to 75 per cent of pre-weaning mortality.

Whenever improvements are made in these two factors, other causes of loss may show themselves, but few, other than disease, will contribute as great a loss as the two above.

OVERLYING

In herds which record the cause of piglet mortality, this is commonly the major reason listed. However, careful consideration should be given to *why* piglets are laid on and detailed records may point the way to a primary problem, which may be little to do with pen design (reckoned to be the major component in this problem).

As the checklist below shows, there are other contributory

factors which should be considered by the pigman before all the blame is heaped upon pen and crate design:

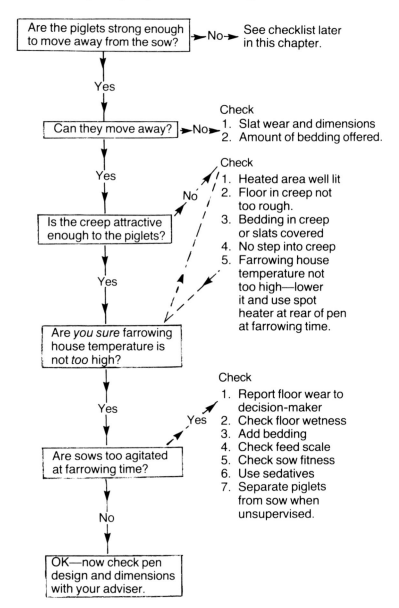

Thus, as can be seen from the checklist, overlying might be recorded as the reason for death, but piglet size, for example, might be the primary cause. Similarly, where sows are grossly underfed around farrowing time they will be more active and the frequency with which they stand up and lie down again will increase the risk of them trampling or lying on piglets.

STARVATION AND ITS PREVENTION

As reductions in the numbers of pigs overlain takes place, the importance of starvation becomes more apparent and, as mentioned, the strength of the piglet in any case will affect the risk that it will be laid on.

The pig's response to efforts to counteract starvation will—in its turn—vary with its size and vigour at birth. So birthweight and vigour must be given high priority in the pigman's mind. Some of the influences on birthweight have been referred to in Chapter 13. The pigman should consider piglet size at birth under three categories, as shown on page 152.

Thus, the need to achieve large litter numbers is slightly at odds with absolute size and variation within a litter but, if the checklist on the right is considered, need not and should not mean unacceptable piglet vigour even though pigs are born with little energy reserve and are poorly developed to cope with low temperatures.

Having achieved viable pigs of, it is hoped, satisfactory size, the pigman must consider his actions to ensure that starvation does not then arise. *The pigman must ensure THAT EVERY PIG RECEIVES COLOSTRUM—preferably from its own mother.*

Colostrum declines rapidly from about six hours after farrowing commences as will the piglets' response to colostrum. Thus, this milk, not only rich in concentrated nutrients, but also in antibodies to protect the piglet, is vital. The pigman can help by:

● ensuring that farrowing is not too protracted—see Chapter 12;
● ensuring that competition does not prevent every pig getting its vital share (this may include removal of some of the bigger, more vigorous earlier-born pigs for a period to give the whole litter opportunity);
● ensuring that the litter dries off quickly, so that they rapidly move and suckle which will help considerably with this initial nutrition. The pigman can help by supervision and suspending a supplementary heater lamp at the rear of the sow during farrowing;

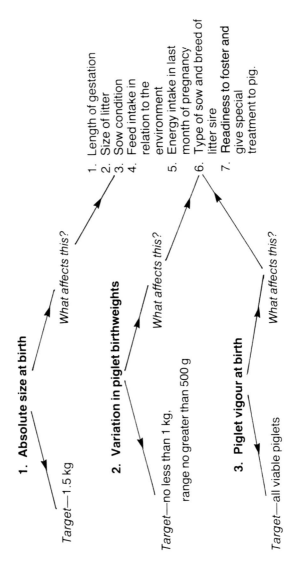

1. Absolute size at birth

Target—1.5 kg

What affects this?

2. Variation in piglet birthweights

Target—no less than 1 kg, range no greater than 500 g

What affects this?

3. Piglet vigour at birth

Target—all viable piglets

What affects this?

1. Length of gestation
2. Size of litter
3. Sow condition
4. Feed intake in relation to the environment
5. Energy intake in last month of pregnancy
6. Type of sow and breed of litter sire
7. Readiness to foster and give special treatment to pig.

● ensuring that colostrum is given where piglets are removed from a sow that may be savage or where cross-fostering takes place.

One of the common problems facing a farrowing house operator is in dealing with pigs of *low viability* at birth. It is especially important that such animals are given priority treatment either by:

● shift suckling (see below)
● fostering (see below)

Even the smallest of pigs at birth are worth efforts to revive them and help them survive. On occasions there may be several weak pigs or even, due to the loss of a sow, orphan pigs, *or* late-born pigs from a protracted farrowing. It may well be advisable to prepare for emergency rearing and boosting of weaker ones as follows:

1. collect colostrum from sows just prior to farrowing, when a flush of milk is apparent, in a sterile container and store in a deep freeze. It is advisable to draw off a small amount of milk from each teat of a number of sows.
2. when needed, warm the milk in a bowl of hot water and prepare equipment ready to administer to the pig by one of the following methods:

(a)
gently open the mouth of the piglet and squirt the colostrum in the mouth using a sterile syringe.

(b)
fill a 20ml syringe with colostrum and 1-2ml corn oil and using a 250-300mm length of silicon tubing which is lubricated with corn oil and gently inserted into the piglet's stomach, allow the mixture to flow down the tube. This technique is best followed initially under the direction of a veterinary surgeon to ensure correct insertion of the tube and sterilisation of equipment.

With either technique it is advisable to administer doses hourly for the first 24 hours, and early treatment raises the likelihood of success.

The pigman must also ensure THAT PIGLETS ARE SAFE-GUARDED AGAINST COMPETITION THROUGHOUT SUCKLING.

It is essential that efforts to ensure survival and even growth are not allowed to slacken once farrowing is over. Initial nutrition is vital but it is as important to shield the individual pig from competition from bigger litter-mates and inadequate milk supply. While aiming at a low mortality to gain an average number of pigs reared per litter of well into double figures it is essential that the pigman accepts that not every sow has the same capability to nourish a litter of that size.

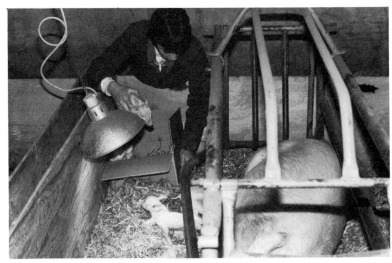

Plate 7. The boxing of piglets can be used to prevent chilling, overlying and savaging and is a useful aid to trouble-free fostering.

Plate 8. Be ready to offer supplementary feed in a clean flat tray to under-privileged pigs.

Quite often, litters which grow fairly evenly for the first week or so begin to show an ever-widening variance in growth rate after this time. The pigman can help with the following procedures.

1. Supplementary Feed
Be ready to administer supplementary feed to the individual. 6–10 ml of warm proprietary glucose product given via a needleless syringe four times per day will assist those piglets falling behind.

2. Liquid Supplements
Offer supplements in liquid form in a flat tray or dish in the creep providing that:

● the utensils are thoroughly cleaned between the three to four daily feeds;
● the material is kept fresh and fed warm;
● particularly for weak pigs (see above), cow's colostrum, if available can be used over the first few days of life following emergency treatment, as described, in order to give strength so that successful subsequent fostering may then be considered.

3. Shift Suckling
Ensure every pig gets its fair share by shift suckling. The litter will be divided into two groups not necessarily of the same number but using individual piglet weight as a guide—ten small pigs suckling for two and half hour periods, matched by seven larger pigs suckling for two-hour periods would help to balance a 'shift suckling' policy. The non-suckling group should be marked and penned in a draught-free area with a temperature of around 30°C and be offered supplementary feed when 'off duty'.

Shift suckling is best used for pigs during their first week of life and gradually piglets will:

● either be fostered off on to other sows as they farrow; *or*
● be weaned earlier on to solid or milk replacement diets.

4. Fostering
Piglets, once they have received colostrum, may be fostered to other sows with piglets at a similar stage or of a similar size. Although the practice of grouping piglets initially by numbers is sound, there must be constant awareness of the need to move round any under-privileged piglets before they become so weak that they die of

'starvation'. In other words, constant regrading of piglets, on a daily basis, must be made over the first ten days or so of life. It is well known that the more forward teats on a sow give more milk and those pigs which select such a teat are at an advantage and grow quicker. Some sows, particularly older ones, milk very poorly on their rear sets of teats due to:

● probable low use earlier in life with declining udder development as the sow ages due to piglet preference for being nearer the 'comfort' of the sow's head;
● increased risk of physical damage near to the hind legs and also from slatted pen areas;
● enlargement of the udder with age which gives reducing access for pigs, particularly the new-born and smaller pig.

For this reason, the pigman should aim, initially, to group piglets on more forward teats, i.e. by size of litter.

Before fostering pigs, there are certain 'ground rules' to consider. As stressed, the sow's milking capacity and ability will vary and the number of pigs which a sow is capable of rearing must be assessed using the following guidelines:

● Has she sufficient teats to accept more pigs?
● Are the teats damaged in any way?
● Are the teats properly exposed?
● Are the teats the correct shape (i.e. not too thick for small pigs)?
● Is the sow fit and healthy?
● Is she of good temperament?
● What is her past rearing record—it is also a good idea to note if any piglets were sub-standard at weaning time as an indicator of the sow's ability to rear a given number of pigs.

Having decided which sow to use as a foster mother, and the number of pigs she might be capable of rearing, the decision about which piglets to foster must be made. For fostering within the first week of life it is normal to:

● Gather the smallest pigs from a number of sows and foster them on to a sow selected on the basis of the seven points above. Observation is necessary to ensure that the small piglets provide adequate suckling stimulus to the foster dam for proper milk let-down to occur.
● This means a readjustment to permit that sow's piglets to be redistributed on other sows according to their size and the other sow's rearing ability.

When cross-fostering is to be used—usually from the time litters are a week or more old, the aim is to reshuffle piglets according to the more obvious milking abilities of the sow which will, by now, be more apparent to the pigman. He can therefore:

- move smaller pigs on to a better milking sow—often as a straight 'swap' between sows;
- consider earlier weaning of the bigger piglets, leaving the smaller animals on to normal weaning age;
- retain a weaned sow on to which smaller pigs are grouped and even given an extended suckling period. With this technique, sow temperament and body condition must be considered as well as the effect upon farrowing frequency if the sow retained for the purpose is not to be culled.

Where fostering and cross-fostering is used, certain other practical points should be contemplated by the pigman:

- every piglet should first receive colostrum from its own mother;
- piglets should be mixed together so that the sow quickly accepts the fostered animals;
- litters showing obvious signs of disease—particularly scours—should not be fostered or have piglets mixed with them from other litters;
- where sub-clinical scours are suspected it is advisable, under veterinary guidance, to treat piglets before cross-fostering to reduce risk of transmission;
- when piglets are fostered back at weaning time because they are considered too small to wean, this should be considered as a failure to respond early enough to the need to give them special treatment. In any case, due to the disruption to the suckling pattern, it takes several days for a fostered piglet to settle into an effective suckling pattern on a foster sow, so one week is unlikely to create sufficient difference in weaning weight and a longer period is required.
- cross-fostering to avoid losses due to starvation and/or to boost weaning is more effective in litters over seven days of age where newly fostered piglets replace a larger piglet weaned or moved onto another sow replacing, in its turn, an even larger pig weaned or moved-on. As stressed, providing scour is prevented, a policy of continuous fostering is the main weapon in reducing starvation losses and increasing weaning weights;
- it is unwise to consider fostering back piglets already weaned and previously moved to weaner pens.

5. Artificial Rearing
Another technique which may be considered, is to rear piglets
artificially. This would be used where:

● there is no suitable foster sow;
● there is too great a surplus of pigs requiring specialist treatment.

Although technology may be envisaged whereby very early
weaning of all piglets may be carried out, the pigman would then be
presented with the equipment and diet for the system. In an
emergency or, on an occasional basis for surplus piglets, the
following routine might be considered.

● The penning and equipment used must be scrupulously clean.
● The temperature must be high—close to piglet's body tempera-
  ture ($37°C$).
● The piglet ideally should receive its mother's colostrum first.
● Regular feeding (every one to two hours), gives best results.
● Cow's colostrum is the best emergency diet and, after three to
  four days this can be diluted, commencing on a 50-50 basis with
  proprietary products or spray-dried whole milk powder.
● Up to 20 ml of liquid per feed may be given depending upon
  frequency of feeding and age and size of pigs being reared.

6. Creep Feeding
A further aid to the prevention of starvation, or the appearance of
the so-called 'nutritional runt', is creep feeding. This will be
discussed in detail later in this chapter.

CHILLING AND ITS PREVENTION

From the previous points stressed, it is now obvious that:

● the smaller the pig, the higher its temperature need;
● the smaller the pig, the lower its energy reserves for combating
  low temperature;
● the lower the temperature, the higher the risk of mortality.

The pigman can help to prevent chilling by:

● ensuring that no sow farrows without creep heating switched on;
● providing a well-lit creep area;
● avoiding too high a room temperature which:
    (a) makes sows distressed at farrowing;
    (b) lowers sow appetite;

(c) increases risk of overlying as piglets are reluctant to move to creep.

But reduce the risk of chilling by:

(a) providing 'roving' lamp at rear of pen when farrowing is expected (care must be taken to ensure that there is adequate load in the building and that the heaters themselves are safely secured);

(b) using a covered creep to create a differential house:creep temperature.

● ensuring that there are no wet patches in the pen because this increases the risk of chilling—pen drainage, clearance of bedding blocking water-flow, maintenance of drinkers and thorough drying of pens before restocking should all be checked.

Farrowing Time Losses and Their Prevention

To ensure proper conduct of all the routines proposed in this chapter and to reduce savaging and suffocation in the membranes (see Chapters 12 and 13), there is no substitute for the supervision of farrowings.

In addition, sows farrowing unobserved may suffer from protracted farrowing and illness which, if not attended to quickly, may result in the loss of the sow as well as her litter.

There is a tendency for sows to farrow during night-time due, probably, to less disturbance at that time. This makes supervision at farrowing less convenient and less likely. However, close supervision of sows, including late evening checks, should ensure that at least 80 per cent of all farrowings are supervised or, at least, monitored at intervals.

Synchronising Farrowings

However, to undertake a more reliable programme of inspections at farrowing time, it may be desirable to consider the synchronisation of farrowings using prostaglandin analogues or by the use of progestagens. These work in different ways:

Prostaglandins—injected intra-muscularly normally trigger farrowing twenty-four hours later. So, by administering the drug one morning, there is a very high likelihood of the sow farrowing during normal working hours one day later. The precision of synchronisation may be increased where an injection of oxytocin is administered 24 hours after the use of prostaglandin.

Progestagens—either injected or added to the feed will delay hormonal change (which causes farrowing to commence), until about thirty hours after the material is last given.

If synchronisation is to be practised, the average length of pregnancy in the herd should be known. Figure 13 shows a typical herd spread and it can be seen that the expectation is for 85 per cent of farrowings to occur on days 114–116 after first service. It is usual for gilts to have slightly longer gestations than sows. Armed with this information, the pigman can be advised by his veterinarian on the use of the products.

There are three considerations that may be made if prostaglandin is used:

1. To inject all sows on the day prior to when more than 10 per cent of farrowings would normally occur (in example day 113), so almost every farrowing will occur on a known day.

2. To attempt to synchronise all farrowings into two predetermined days in a week.

3. To avoid weekend farrowings by injecting on Thursday any sows due to farrow on Friday, Saturday or Sunday and any gilts due on Friday or Saturday.

General advice is that:

1. The earlier farrowing takes place, the greater the risk of smaller pigs being born (see notes on small piglets earlier in this chapter).

2. The products are best used to group together farrowings on to predetermined days to ensure close supervision is applied and, in any case, bringing forward farrowings by more than two days increases problems of piglet survival even with close post-farrowing supervision.

Using prostaglandin

Because the injection is small (usually only 2 ml per sow), it is vital that it be accurately given. Thus, ensure that:

● the needle is long enough (16 gauge × 40 mm needle);
● the injection is located intra-muscularly – preferably in the rump;
● there is no leakage from the injection site (ham or neck);
● a record is carefully made of drug use.

Using progestagens

If these are the feed-additive type, add to feed from the day prior to the earliest likely farrowing date and withdraw thirty hours before farrowing is planned.

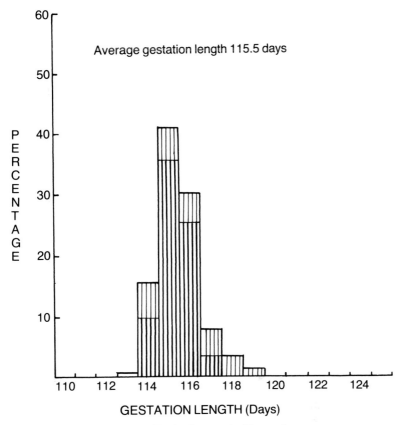

Fig. 13. Typical spread of farrowings

Losses at farrowing may arise from:
● suffocation due to prolonged farrowing (later-born piglets tend to be at greater risk due to premature separation of navel cords);
● piglets enclosed in membranes;
● protracted farrowings (see below);
● crushing by an agitated, restless or clumsy sow;
● chilling;
● being very small at birth, in which case supplementary doses of warm glucose may be given;
● sow illness due to farrowing fever or stress;
● savaging (see Chapter 13);

- early competition (see fostering and shift suckling), but remember that any piglet born early must be given special attention;
- navel bleeding (see below).

When and How to Assist Farrowings
One of the disadvantages of supervised farrowings is that there is an increased temptation to examine the sow internally, whereas this should only be carried out in an emergency and should not be necessary in more than 10 per cent of farrowings. It is not a simple judgement to decide upon the need for assistance, but the following checklist may be used as a guide to intervention.

1. Are you sure of the farrowing date?
2. What do the sow's past records show?
3. Are all normal farrowing signs obvious:
   - slackening of muscles;
   - swelling of vulva;
   - presence of milk;
   - expulsion of blood;
   - 'nesting' behaviour;
   - straining and pawing of hind leg?
4. Has there been no birth for thirty minutes?
5. Is the sow still straining?

If all the above checks have been made and the pigman is satisfied that a potential problem exists, the following routine should be followed:

1. Wash hands/arm thoroughly and scrub finger nails.
2. Wash sow's vulva.
3. Lubricate hand/arm with antiseptic gel or soap.
4. Insert hand carefully along vaginal passage and check for obstruction around cervix.
5. Enter womb carefully and attempt to locate constriction or piglet or cleansing.
6. After examination, insert uterine pessary or administer antibiotic to prevent infection.

The use of drugs (oxytocin) to speed contractions of the womb and milk let-down may be given under veterinary supervision as well as antibiotics to prevent infection following assistance. Proprietary uterine relaxant drugs may be used instead of oxytocin to speed farrowings; one advantage of this is that they have less stimulus upon the milk let-down process.

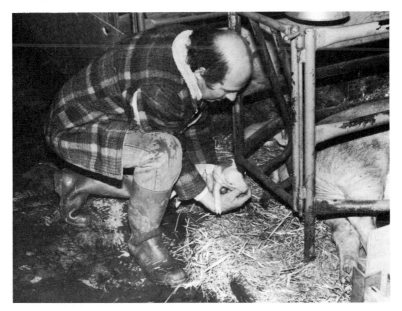

Plate 9. Attendance at farrowing can be a useful way of reducing losses.

Reviving New-born Piglets

One of the greatest rewards for the effort involved in supervising farrowings is being able to revive pigs which would not have survived without the intervention of the pigman. These include:

- clearing the mouth and nose of mucus;
- removing membranes from around head;
- applying kiss of life to non-respirating piglets;
- 'shock' to encourage piglets to breathe by slapping or plunging in cold water;
- rapid drying of the chilled or weak piglet.

Navel Bleeding

The causes of navel bleeding are not clear, but the problem must be dealt with very rapidly because the pig's reserves are small and if unchecked, survival will not extend beyond a few hours. *Bleeding cords must be clamped with navel clips or tied up to prevent haemorrhaging.*

It is then advisable to administer iron injection to pigs immediately to assist in restoring haemoglobin levels.

The following points should be checked by the pigman:

1. Is the farrowing protracted? Later-born pigs are more likely to bleed.

2. Is the farrowing premature? Induced farrowings may increase navel bleeding.

3. Are the floors rough? Can bedding be added?

4. Are cords trapped in slats? Observe farrowing.

5. Are sharp-edged shavings used? Use chopped straw.

6. Is sow fed mouldy or contaminated feed? Mycotoxins may interfere with vitamin uptake and it is suspected that both vitamins K and C affect the incidence of navel bleeding.

7. Have disinfectant residues been washed from surfaces?

Even where navel bleeding is not a problem, it may be advisable to dress or dip the cords in an iodine solution to speed their drying to reduce risks of bacterial entry.

Splayed legs

Splayed-legged piglets may arise from a combination of factors. It should rarely be seen in more than 0.5 per cent of farrowings, but indoor units often report an outbreak of the problem, followed by a return to normality which is equally difficult to explain. Although genetics may play a part, it has a major environmental relationship and is rarely recorded in outdoor units.

Slippery floors and slats do not help splayed legs, but they do aid hygiene, so observation of farrowing helps to mitigate the condition. Some suspicion exists that mycotoxins in the feed of sows and reproductive virus infection may also have a part to play in the occurrence of this condition.

The pigman can help by:

● taping affected legs together to speed recovery which normally takes place within a few days;
● ensuring affected piglets get their share of colostrum and milk;
● assisting piglets with splayed legs to avoid being trampled or overlaid during the first few days of life;
● adding shavings or chopped straw.

ROUTINE STOCK TASKS

At birth, or within a few days of it, a group of operations are necessary. A description of the conduct of each task will not be given, but the need for attention to hygiene in all such operations must be stressed.

Weighing of litters
This is valuable when attempting to determine:

● the status quo;
● effects of any husbandry changes;
● accuracy of batching for fostering.

It is important that an easily cleaned container be used—plastics are ideal. If not regularly cleaned such a container may act as a reservoir of infection.

Teeth clipping
This helps to reduce:

● sow discomfort and, indirectly, mastitis;
● damage to piglets' faces and, indirectly, infections.

It is important that the pigman remembers that the mouth is a source of most bacteria, so the teeth-clippers should be sterilised between litters and never used for other tasks such as tail docking.

Castration
If this unpleasant task has to be carried out, it is best done early in life to minimise stress on the piglet.
Hygiene is an important consideration and treatment of the wound with an antiseptic dressing is essential.

Tail docking
This is a useful means of controlling tail-biting in growing and finishing stages. This may be outlawed by legislation in the future.
Where practised, it is necessary to:

● use an implement which is separate from that used for teeth clipping—preferably a cauterising tool as used for de-tailing dogs;
● ensure proper hygiene;
● remove over half the tail if it is to be a really effective deterrent to tail-biting.

Identification
It is vital to practise an ear-marking code, so that finishing pig performance can be monitored (see later chapters).

Notching or tattooing may be considered according to a pre-set system for determining age at least, although sire and dam identification may also be added, particularly in a pedigree herd.

It is vital where replacement breeding gilts are reared on the farm to give each gilt its individual identification, so that no confusions exist should fostering take place.

Recording

It is vital for future judgements on management modifications to record accurately all actions including treatments and fosterings.

OTHER SUCKLING PROBLEMS

There is little that the pigman can do if trembling piglets or blind anus is observed. Trembling pigs may recover and the cause is undoubtedly attributable to some interference during pregnancy, and Aujeszky's disease is often suspected if the problem is widespread and accompanied by other symptoms.

Male pigs usually bloat and die if affected with blind anus, but gilts may survive with a single orifice.

Joint Ill

This is a particular problem on some units, and may occur in spasms. Usually bacterial following entry of infection via:

1. *the navel at birth*—dress with iodine;
2. *a wound*—clip teeth;
3. *tail docking*—use separate tools and practise strict hygiene.

Damaged Knees

A very common occurrence which may also be associated with the loss of anterior teats (teat necrosis). The pigman may help to reduce this problem by:

● ensuring piglets do not have to work too hard for their milk—fostering and eliminating sows with enlarged udders may help;
● providing bedding to mask effect of floors;
● checking floors and slats for damaged surfaces and edges;
● ensuring no residue of disinfectant remains on floors after cleaning pen.

Scour

On some units, or at particular times, this can be a particular problem and the cause of extensive loss. Scour may be apparent soon after birth (neo-natal) or some time later. The most probable cause of piglet scour is a build-up of *E. coli* bacteria, although other agents such as clostridia and coccidiosis may be implicated. Whatever the cause the pigman has a large role to play in preventing the disease initially and in its control should an outbreak occur. It is too simple to resort to sow vaccination without reviewing husbandry routines. A breakdown or variance in any of the points listed – due to disruption by herd expansion or even relaxation in thoroughness of routines – can trigger and cause spread of scours.

Although it is advisable to check all husbandry points first, a pigman should not be slow to discuss sow vaccination or the use of antibiotics prior to farrowing where the problem persists.

Scour outbreaks can frequently be traced to an important change in unit operation. Peaks in farrowing schedules, a large influx of gilts, breakdown of pressure washer, a change in the weather may act as a trigger to bacterial build-up and a scour outbreak.

The pigman can play his part by:

● reducing the burden of bacteria by never omitting important hygiene measures;
● helping to raise herd immunity;
● seeking veterinary help promptly to limit the spread of any problem observed.

Although it may be possible to incorporate *E. coli*-resistant strains of stock in the future, it will still be necessary for the pigman to practise the following routines to prevent clostridial infections and other bacterial gut invaders.

1. Follow hygiene routines precisely (see Chapter 6).
2. Ensure that farrowings (and therefore services) are regulated to allow all in/all out policy and proper cleaning-out routine to be practised.
3. Wash sows thoroughly prior to farrowing.
4. Prevent chilling of pigs – attendance at farrowing and checking of pen temperature will help to overcome this, *the major cause of piglet scour.*
5. Keep separate tools in each farrowing house or room.
6. Dip feet in fresh disinfectant when passing from house to house.
7. Wash hands after handling sick pigs.
8. Keep non-farrowing house personnel out of farrowing rooms.
9. Do not foster scouring piglets back on to other sows.

Scours which occur in the first few hours of life need additional responses to help build up antibody level in the mother. The problem is worse in gilts, but the following may even be practised in sows when there is an outbreak of scour on any unit.

1. Feed back scour to gilts or sows in their last month of pregnancy to boost antibody production.
2. Vaccinate sows using proprietary products, but these only work if:
   ● vaccination is carefully carried out according to directions;
   ● it *is E. coli*;
   ● the vaccine contains antiserum to the strain of *E. coli* involved.
3. Then ensure that the litter receives colostrum.
4. The pig tasks referred to may be best delayed until the litter is well established.

As stressed, where an outbreak does occur, it is essential to:

● obtain rapid diagnosis;
● treat (orally dose), affected and in-contact pigs promptly and for as long as is prescribed;
● practise strict hygiene to reduce the risk of spread to other pigs.

Atrophic Rhinitis
Where this condition occurs on a unit, care must be taken to:

● minimise the mixing of pigs of differing ages, particularly up to eight weeks of age;
● treat all piglets with antibiotic or vaccine in the first four weeks of life to break the infective cycle which commences with the sow;
● ensure adequate air exchange occurs without draught or wide fluctuations in temperature.

CREEP FEEDING

The value of creep feeding in reducing loss of sow condition during suckling and increasing weaning weights of piglets was unchallenged until weaning ages started to be reduced below four weeks. There are among early weaners those who consider the time needed to carry out creep feeding and the extra burden on cleaning utensils unrewarding.

However, it is possible to:

● increase weaning weights even of early-weaned piglets;
● reduce post-weaning stress because pigs are bigger *and* used to solid feed.

The following notes are suggested as a blueprint for a creep-feeding routine.

1. Always use fresh feed—never store creep in the farrowing house, always re-seal the bag after use, always order fresh creep every week.

2. Never offer more than 20 g per litter per day until they are obviously eating it.

3. Always remove yesterday's creep feed and offer fresh.

4. Never use a hopper until pigs are *four* weeks old.

5. Feed on the floor or on a clean, flat receptacle.

6. Never offer creep within two hours of the sow being fed as the litter will:

- be waiting to suckle, or
- will be suckling, or
- sleeping after suckling.

Early uptake of creep will only occur if the pigs are tempted during their 'active phase' following the big and deep suckles the sow gives them after her own feed and then only from five to six days after birth onwards.

7. Water is of no consequence for creep uptake if pigs are weaned under four weeks of age; however, water provision will help to overcome dehydration problems.

There is a debate about the merit of creep feeding at all for pigs weaned at under one month of age. Some operators report a decrease in post-weaning diarrhoea where creep is *not* offered prior to weaning. However, this applies only when insufficient creep has been consumed to actively prime the enzyme system in the pig's gut to aid digestion of the solid feed. So it can be fairly said that where creep feeding is carried out badly and pre-weaning intakes are less than 650g/pig before weaning, then it may be safer not to creep feed at all. The corollary of this is probably more appropriate – that is to creep well so that all pigs have eaten good quantities of creep prior to weaning. This will not only help to prevent post-weaning scours but will speed the time when ad lib feeding can be used and so increase growth in the period after weaning.

WEANING TIME MANAGEMENT

It is now acknowledged that pigmen should avoid weaning weak piglets or those so small that post-weaning performance is likely to be compromised. As mentioned under the section on fostering the need to back-foster more than 4-5% of piglets because of this problem suggests that attention to balance weaning weights has not been good enough or action taken early enough.

Where problems are caused by piglets being weaned too small, varying techniques may be considered:

● Staggered or 'cascade' weaning: This is where piglets of the desired weight and vigour are progressively removed from the sow, leaving smaller pigs behind until they in turn are large enough. Whilst this is useful in creating more even piglets at weaning it can create problems in that it may be more difficult to follow an all-in/all-out hygiene routine for farrowing houses.
● Back-fostering: This can be used in conjunction with the staggered technique where larger pigs are weaned and replaced by smaller piglets from another sow or other sows, leaving just one sow behind to act as a foster sow. (See also page 155.)
● Exchanging sows: This technique can be used on its own – simply by weaning one strong, probably younger litter and moving into that pen a litter of smaller or even back-fostered pigs. This enables pens to be used more efficiently and hygiene routines to be unhindered. It can also be used in conjunction with the other two weaning techniques described.

The delay of weaning may also have secondary effects, apart from a potential disruption of between-batch cleaning routines, and this is to influence the average weaning age in the herd or even a distortion of the planned service programme. If a sow is to be used as a foster mother for piglets too small for successful weaning, then it may be desirable to choose a sow due to be culled, providing a suitable animal is available, and to foster onto her no more piglets than she was previously suckling.

Remember
Attention at farrowing time and up to weaning can be measured in more (or less), bigger (or smaller) pigs and drug bills.

Immediate attention to deal with sows in difficulty at or around farrowing time and whenever an individual does not eat a normal allowance is usually rewarded by a quicker, and less expensive, recovery. Similarly early action to deal with under-privileged piglets and continuous efforts to keep piglets warm at all times, particularly in the first week of life, will reduce problems with mortality and help to prevent debilitating scours which also influence post-weaning performance.

SECTION VII

Chapters 15–16

Weaner and Grower Husbandry

Including checklists for:

Improving weaning weights

Beating post-weaning checks

Feeding the newly-weaned piglet

Overcoming poor post-weaning
growth rates

Penning considerations

Meningitis, 'mulberry heart' disease, pneumonia, rhinitis, mange,
'greasy pig' disease, navel sucking.

Managing the bought-in weaner

Body tissue development

Feeding the growing pig

Water system checks

Prolapse, 'bent-backs', tail biting, ear sucking, fighting.

Chapter 15

A few, quick facts—weaning is a major stress hurdle which will only be cleared with the aid of the pigman who must ensure that:

house temperature ⎫
dietary intake ⎪ are within the range with which the pigs,
disease challenge ⎬ whatever their age and size, can cope.
competitive problems ⎭

The smaller the pigs are at weaning, the more specific are their needs for all these conditions. Thus, early weaning, say under four weeks or 8 kg, demands more of the environment and the pigman. In many instances the pigman is faced with weaner accommodation which was designed for pigs of a certain weaning weight but must now be used for pigs which are weaned earlier. Either that or the pigs are simply not as big as they should be at weaning.

In such circumstances, the pigman should:

● review the recommendations given in Chapter 7 on temperature adjustments and checks;
● reconsider pre-weaning management.

If there is a need to review pre-weaning management, the following influences upon weaning weight should be considered:

1. Is birthweight satisfactory? (See Chapter 14.)
2. Are sow condition and sow feeding satisfactory to encourage milk yield? (See Chapter 14.)
3. Is sow feed contaminated, so depressing milk yield?
4. Could better average weaning weights be achieved by fostering? (See Chapter 14.)
5. Would supplementary feeding help? (See below.)
6. Is creep feeding technique good enough? (See Chapter 14.)
7. Could weaning be delayed without compromising farrowing routines or results?
8. Is it practicable to delay weaning those pigs which are below a minimum weight-for-age at weaning by using sows which are due to be culled as foster sows?
9. Is it possible to create a specialised pen into which pigs which

173

are really too small for the normal housing system are placed for a week or so after weaning?

If it is not possible to buffer the weaners satisfactorily against a suspect environment by any of the above means, or if there are occasional 'problem' pigs, the pigman must consider giving the pigs some specialist treatment. This is important not just in order to help pigs:

- to make up lost ground, *or*
- to be able to cope better with the adverse environment, *but also*
- to prevent them slipping further behind and, perhaps, not surviving;
- offsetting the problems associated with scours.

A major key to preventing these associated problems is to increase the piglets' liquid intake and to replace body fluids lost as a result of scour. It is likely that, in the future, pigs weaned earlier than three weeks of age will be fed a liquid diet for a period after weaning because liquid feed undoubtedly enhances the early-life performance of pigs. Because of this the pigman should:

- provide water plus sugars and essential minerals to offset losses which cause the pig to become dehydrated (dehydration is a common cause of death due to a slowing down of the body processes and, although it is important to tackle the root causes of infection, offering a liquid fortified with a proprietary electrolytic/sugar solution can have a considerable advantage);
- ensure that all equipment used is cleaned and the solution fresh and uncontaminated;
- ensure that the technique can be and *should* be used as soon as a litter or pen of pigs are seen to be losing ground on their contemporaries—both before and after weaning.

Although infective agents like *E. coli*, other bacteria and viruses, may be diagnosed from scouring animals it is important to remember the following points.

1. These organisms are probably present on every pig farm.
2. That a comprehensive hygiene programme will help to control the burden.
3. That there is a need for some management action or a failure to offset the effects of some change to enable the bacteria or virus to multiply and become a problem.
4. As soon as the pigs' intestines react to an unfavourable set of circumstances, the conditions in the gut change so that the small and large intestines become filled with part-digested feed. Thus the

pig's body tries to get rid of these products and to bathe the gut lining by now damaged by toxins produced by bacteria or viruses 'breeding' on the undigested feed. This is the sequence which leads to the pigman recognising scour in the weaner and the need for replacing those liquids which turn dung into a multi-coloured diarrhoea or even cause expansion of the gut which leads to bowel oedema when the pig collapses and dies.

AVOIDING POST-WEANING CHECKS

Between weaning and 20 kg little pigs should gain between 400 and 450 g per day (just under 1 lb a day). Apart from aiming to improve weaning weight and dealing with poorer pigs, the pigman must consider the following checklist.

House Temperature

DO NOT	*BUT DO*
1. Presume that the temperature set on the controller is providing that climate for the pigs	*Check it by:* (a) Lying patterns of the pigs. (b) Accurate recording equipment.
2. Presume that air temperature in the house is repeated at pig level.	*Check for:* (a) Draughts—these reduce the animals' tolerance of low temperatures. (b) Adverse air current.
3. Assume that a setting for the average or largest pig is good enough.	*Set temperature for:* (a) The smallest pig in the pen because the smallest is the most susceptible and, up to 10 kg, a temperature slightly too high will have no adverse effect on the fitter pig.
4. Assume that the temperature can never be too high.	*Modify temperature so that:* (a) It is adequate to prevent check initially. (b) It is reduced to offset depressed appetite as the pigs grow.
5. Assume that there is nothing that can be done to offset low temperatures.	*Try to:* (a) Offset disadvantages by compensating with improved nutrient intake (higher feed scale or better diet). (b) Grow pigs to a heavier weight before moving them into the weaner building.

The Feeding System

DO NOT:	BUT DO:
1. Overload the gut with excess feed.	*Try to feed as much as possible by:* (a) Acclimatising the gut to the weaner diet by good creep feed management before weaning. (b) Avoid a diet change around weaning time.
2. Rule out ad lib feeding.	*Consider the:* (a) Diet used and points (a) and (b) above. (b) House temperature (see previous table). (c) Hopper design and feed freshness.
3. Assume all the pigs are eating what you think.	*Check that:* (a) There is adequate feed space per pig: 150 mm for ad lib 200 mm for controlled feeding. (b) Automatic equipment is accurate. (c) Feed is fresh.
4. Assume that the water system is satisfactory.	*Check the:* (a) Effect on appetite that water has and use the checklist in Chapter 16.
5. Assume that medicines are not working.	*Make sure that:* (a) Drug inclusion is at the rate recommended by the veterinarian. (b) Pigs are offered and are eating sufficient to obtain full benefits of the drug.

The Penning System

DO NOT:	BUT DO:
1. Overcrowd the pens.	*Check that:* (a) Pigs have 0.3 square metre floor space up to 20 kg. (b) Stocking rate is not so low that a satisfactory temperature cannot be maintained.

DO NOT:	*BUT DO:*
2. Put too many pigs together for the facilities provided.	*Is the:* (a) Feed space allowance adequate? (see above list). (b) Watering facility suitable? (see Chapter 16). (c) Low number (maximum twelve) per pen seems to give a performance edge up to eight weeks of age.
3. Mix age ranges in a building.	*Try to:* (a) Have pigs born within two weeks, and preferably one week of each other, in same airspace. (b) Avoid holding back a reservoir of 'poor doers' which might 'infect' incoming pigs.
4. Mix weights within a pen.	*The aim should be:* (a) Keep pigs within a weight range of 2 kg at this stage. (b) Always put smallest pigs together. (c) Group by size rather than sex at this stage. (d) Try to wean even-sized litters so they can be penned together at weaning.

There is little doubt that mixing presents many different traumas for the newly-weaned pig and that if pens are not too large (which the pigman cannot easily influence), and litters are even-sized when weaned (which the pigman *can* influence), the ideal weaning arrangement is the penning of pigs in litter groups.

OTHER WEANING-TIME PROBLEMS

The other causes of difficulty for the pigman at weaning time are losses and performance shortfalls caused by a range of conditions.

These, apart from scours and bowel oedema (which can be confused with meningitis), dealt with earlier in this chapter, may be categorised as:

1. Those affecting nerves or heart.
2. Those affecting the respiratory system.
3. Those affecting the skin.

All such problems require positive diagnosis, so close contact with the unit veterinarian is called for. However, like all health conditions, the pigman can contribute by:

● a rigid preventive programme.
● prompt action if the conditions are observed.

1. Diseases Affecting Nerves or Heart
The problems with such conditions is the speed with which death follows if early treatment is not given.

CONDITION	PIGMAN'S CHECKLIST FOR ACTION
1. Meningitis Little prevention is possible due to the method of spread which is via carrier sows or infection with haemophilus.	(a) Promptly treat affected pigs, mark and repeat treatment twice daily until pig has had minimum of three days medication. (b) Separate from pen-mates to avoid damage. (c) Gently and carefully administer electrolyte solution to reduce weight losses and dehydration. (d) If a major problem, veterinarian may advise herd, rather than individual, treatment.
2. Mulberry Heart Possibly genetic, but is masked by high vitamin E and selenium levels in diet.	(a) Watch carefully for pigs growing very rapidly or those suddenly responding to an improvement in their circumstances following a growth check. (b) Because high energy diets, which incorporate oils and fats, are associated with this faster growth check the vitamin E and selenium levels.

2. Diseases Affecting the Respiratory System
Enzootic pneumonia is one of the most common pig conditions throughout the world and atrophic rhinitis is another common scourge.

CONDITION	PIGMAN'S CHECKLIST FOR ACTION
1. Enzootic (mycoplasma) pneumonia The pigman can help to minimise the effects of this condition but may need the help of medicated diet or water supply if crises occur.	(a) Ideally all-in/all-out policy will be adopted throughout the unit to reduce infection of younger, more susceptible pigs. (b) Keeping pigs of limited weight range in house means that the environment can be matched more precisely for them. (c) Adjustments to vents and control equipment have a large influence on the incidence of this disease, especially in spring and autumn and to reduce dust and other irritants.

CONDITION	PIGMAN'S CHECKLIST FOR ACTION
	(d) Avoid over-stocking of pens.
	(e) All incoming stock should be carefully acclimatised.
	(f) Special care must be taken when the average age of the herd falls due to the lower immunity levels of gilts and the effect of this on their litters.
2. Atrophic rhinitis Requires close veterinary supervision.	(a) The checklist above is important for rhinitis also.
	(b) In addition, vaccination of the sow and medication for suckling pigs may be required to minimise the problem at weaning time (see also Chapter 14 and previous notes on pennings).
3. Haemophilus Differing strains, one of which can cause severe mortality. Distinctive 'barking' cough can be heard.	(a) In addition to above notes for other respiratory conditions, it is vital to spot and quickly treat affected animals as death may be rapid.
	(b) Haemophilus may affect pigs of *all* ages and has a much quicker onset than enzootic (mycoplasma) pneumonia with pigs becoming affected within a few days.
4. Coronavirus Often quite transitory but can leave a long legacy of lost performance.	(a) Thought to be wind transmitted and respiratory distress and raised temperatures caused can have far-reached effects on many aspects of herd performance.
5. Swine Influenza Rapidly spread virus.	(a) Easily confused with coronavirus and has similar effects.
	(b) Conditions summarised under points 3-5 normally necessitate herd medication to reduce effects and speed recovery.
6. P.R.R.S. ('blue ear') Rapidly spread virus.	Immediate effect is to pre-dispose animal to other conditions and secondary infections, so good husbandry and rapid treatment (as described under Enzootic pneumonia) are essential where this virus is endemic within a herd.

3. Conditions Affecting the Skin

Both the conditions below are referred to in chapters on breeding stock management. However the pigman responsible for weaner pigs would consider the following:

CONDITION	PIGMAN'S CHECKLIST FOR ACTION
1. Mange Spread by a parasite.	(a) All weaners should be dipped in a warm mange solution at weaning time.
	(b) Any obviously seriously affected pigs should be scrubbed before dipping.
	(c) In persistent cases, the pigs should be tightly penned and sprayed/treated again two weeks following weaning.

	(d) Pens which have held affected stock should be thoroughly cleaned before re-stocking.
2. Greasy Pig Disease Spread by carrier pigs.	(a) The above actions should be repeated.
	(b) Especial attention to the air movement system should be made with particular attention to humidity levels. If too humid, so there is obvious condensation present, the problem may be worse.
	(c) Probable carrier sows in the herd should be identified and treated by scrubbing with an antiseptic solution.

Navel Sucking

This is a condition predominantly associated with pigs weaned relatively early in life—at three weeks, or less. In most cases the pigs most likely to be involved are underprivileged pigs, particularly those which are penned with larger pigs after weaning. The pigman can help to reduce this problem by:

● Pre-weaning management, particularly by attempting to even-up litter groups by cross-fostering and supplementary feeding.
● Removing smaller pigs from a litter and penning with other pigs of a similar size.
● Providing supplementary glucose or electrolytic solution after weaning.

OVERCOMING POOR GROWTH IN THE POST-WEANING STAGE

It is in the first month or so after weaning that the biggest discrepancy between potential and achieved growth rate typically occurs. Piglets can be expected to gain possibly only 150g/day on average over the first 5-7 days after weaning but this should double in the following week and average over 400g/day over the first month.

This statement of target growth immediately highlights the root cause of one major reason for poor growth – that the pigs are not offered or, for other reasons, do not consume sufficient feed.

Pigs weaned in the first month of life have extraordinarily efficient feed conversion potential helped by the very low body maintenance requirements at that stage. Thus, the intake of 120g/day will yield only 120g of growth at best *yet* many operators insist on restricting intake for some days after weaning and continue to over-restrict and partly *starve* the piglets beyond the point where post-weaning stress might be reasonably expected to occur. The need to restrict normally follows some experience of post-weaning scour but a number of factors can be altered to reduce the risk of scours and, so, to allow more generous feeding to permit normal growth.

The checklist below shows how the anxiety concerning a scour outbreak might be reduced and other ways in which a pigman might ensure safe but more generous levels of feed are consumed to overcome poor growth in the post-weaning phase:

- Avoid weaning piglets that are too small by back-fostering or exchanging for bigger piglets from younger litters.
- Ensure that pen temperature is suited to size of piglet weaned; unheated kennel types can often be maintained at an appropriate temperature only at the expense of poor ventilation and over-stocking which, in themselves, slow down growth and increase the likelihood of disease.
- Always try to follow a strict all-in/all-out routine with thorough hygiene.
- Try to encourage piglets to eat generous quantities of creep before weaning (see Chapter 14).
- Ensure that the creep is of a quality that suits small pigs, with particular care taken with the quality of raw materials used.
- Avoid pellets which are too large or excessively hard.
- Consider feeding 'little and often' to increase freshness and interest in the feed.
- Ensure that the feed container is easily accessible: the lip over which newly-weaned piglets feed should not exceed 100mm in height and each divider should be at least 80mm wide.
- Whilst floor feeding depresses feed uptake and is to be avoided wherever possible, where it *is* used there is a need to offer sufficient feed to avoid under-feeding yet avoid excessive wastage due to spoilage. As this is an extremely difficult balance to achieve, trough or hopper feeding is strongly preferred.
- Ensure that water is available in a condition that piglets relish; avoid excessively cold temperature and drinkers that may be difficult to operate.
- Do not store feed in the weaner building where it can become stale and unpalatable prior to being offered to the pigs.

Remember
That the 'rules' for successful weaner pig management are the same regardless of weaning age, but they become more important the smaller the pigs are at weaning time.

Chapter 16

Due to the difficulties which surround the weaning period, there seems to be a practice among many pigmen to relax the supervision once this tricky phase has been negotiated. Industry performance figures, and those from individual herds, suggest that pigs in their third and fourth months of life rarely achieve their full potential for growth. This is a little more understandable in the case of the specialist fattener who probably buys in his pigs in their third month of life.

A study of the way in which the pig's body tissue develops shows that it really is important that full growth is achieved in the first three months or so of life. This is because it is in this early period that muscle is developed more efficiently than fat—if we fail to capitalise on this fact there is no chance of recovering the opportunity to put on that lean later. So there is every likelihood that, if pigs are restricted in early life growth, they will either:

1. Have to grow fast later to catch up, which will mean:
● More fat than a pig grown quicker in early life; therefore, poorer grading.
● Because it takes more feed to gain fat than lean, the feed conversion ratio will be poorer.
2. Alternatively, the pig will be older when it reaches slaughter weight, thus:
● Unit throughput may be slower, leading to overstocking of pens or the selling of lighter pigs.
● Due to the fact that the pig lives longer, it will need more feed for maintenance, or simple needs of living, again making feed conversion worse.

Although there are a wide number of reasons why pigs grow at different rates and produce different degrees of fatness (see Chapter 17), it is a sound rule that every effort should be made to capitalise on the pig's inherent capability to gain lean meat in a more efficient ratio to fat earlier in life. Study fig. 14 and you will see how a failure to capitalise upon the early-life potential of a pig can lead to the production of an animal with a very different whole lifetime performance.

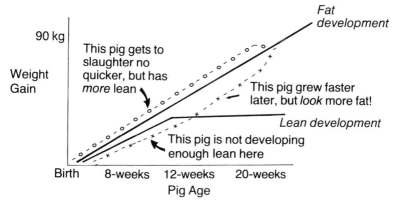

Fig. 14. Feeding for leanness

It must be stressed again, that a failure to achieve that desired pattern of growth is only *one* reason for poor finishing herd performance. Nevertheless, it is one that cannot be recovered by later management efforts. It guides the pigman to a *feeding strategy* which suggests:

● *maximum growth in the first few months of life;*
● *later life growth controlled to provide the carcase quality demanded.*

All too often the pigman is faced with pigs which have grown less well early in life, so he has to try and recover this situation by feeding more in later life with the unsatisfactory result shown in fig. 14.

So, as a general principle, the pigman should use this grower period of the pig's life to set the standard for lifetime performance.

THE BREEDER/FEEDER UNIT

There will be occasions when the pigman sees a need to control feed intake in the post-weaning phase—if there is an outbreak of scour for example—but the aim should always be *to achieve an average eight-week weight of 20 kg with less than 10 per cent of all pigs below 18 kg.*

If this cannot be achieved, then every effort should be made to make up the lost ground in the next few weeks of life, and not to allow pigs to drift further and further behind as time passes.

This demands that careful consideration be given to the likely effects of this upon lifetime pig performance. Where pigs are small for their age:

● they need a better temperature and environment;
● they need a better diet to allow them to take in more nutrients in a small quantity in order to catch up;
● if they have to share an airspace with bigger pigs and do not receive more nutrients then they will slip even further behind and may even struggle to survive.

Just as it is inadvisable to wean pigs that are much smaller than the rest, so it is unwise to move weak pigs from a weaner to follow-on house. This may mean an exchange of a younger but bigger pen of pigs for a pen of smaller ones or even the swapping of individual pigs if this can be achieved without fighting.

There is, therefore, a need to treat the below-average pig as a 'special' at this early stage. This is a sound reason for ignoring average pig growth rates and tackling growing pig management as a need to correct *below average* pigs. This produces the same result as raising the average in any case, but to greater effect because it will almost always be these 'below-par' pigs which struggle to survive or, at very least, end up occupying a pen long after the main group has gone.

So, for performance reasons *and* ease of pigmanship, it is vital to achieve good early life performance.

PURCHASED WEANERS

There has been a gradual evolution within the pig industry towards integrated pig enterprises but it is estimated that one pig in three is still finished on a farm different from the one on which it is born.

This poses an additional problem for the pigman who has to finish such pigs because:

● transport in itself is a form of stress;
● it might well mean pigs from more than one source being mixed;
● he has little control over their management up to this point.

Although purchase of weaners from a single source reduces the risks involved, it is still advisable that:

1. The reception area for the pigs is completely separate from the main finishing population.
2. Close observation be made to determine quickly the:
● need for diet medication to overcome stress-related con-

ditions and to reduce the risk of transmission of incoming disease to the resident herd;
● need to treat all pigs against internal parasites and skin disease on arrival or before movement into the main herd.

The principal requirements for the introduction of stock into the herd will not differ, except in detail, from the recommendations given for the acclimatisation of breeding stock (see Chapter 10). Clearly the pigman will not need to consider reproductive viruses. but concern over respiratory diseases and scour will be heightened.

Practical Steps to Reduce Health Problems
Although these measures are related to purchased weaners, the techniques are equally applicable to home-reared pigs. The pigman can help to minimise the effects of pneumonia and rhinitis by ensuring that the following actions are taken.

1. Reduce the age and weight range of pigs within an air space.
2. Do not mix incoming pigs with others on the unit within one month of arrival.
3. Ensure that pigs are provided with a satisfactory temperature in the sleeping area (over 20°C at *all* times).
4. See that the temperature does not fluctuate between day and night within the sleeping area (not more than ± 1°C and *preferably* ±0.5°C).
5. Ensure that any pigs more than 2 kg below the average of the group are separated and given:
● an even better environment;
● less competition;
● a better diet;
● medication, if necessary.
6. Any pigs showing signs of ill health should be immediately withdrawn and medicated to the recommendations given by the veterinarian.
7. Ensure that the group size be kept as small as possible—below twenty pigs per pen is preferable.
8. See that no slow-growing pigs are left behind to act as a potential 'reservoir' of infection for the incoming stock.
9. Do not mix pigs with the resident herd until they are large enough to cope with:
● the disease challenge;
● the environment compromise.
This is likely to be at around 60 kg or when pigs are moved to the finishing house.

10. The reception area must be thoroughly washed, disinfected and rested when the batch of pigs has been moved out. This is of particular importance in the control of gut diseases, although all-in/all-out policies are also a prime weapon in the pigman's armoury in coping with pneumonia and rhinitis. Disinfection foot baths between new and resident stock will help to reduce the risk of contamination.

11. Consider the use of strategic medication. Because of the stress involved and the unknown health status of supplying herds it may be necessary to medicate all pigs on transfer to underwrite the hygiene programme. Pigs do not only carry organisms ready to spread lung and snout diseases, but bacteria responsible for bowel disorders, including *swine dysentery*, or 'bloody gut'. The sensitivity of dysentery-causing organisms to drying underlines the need for good hygiene and drying of pens between batches, and the breaking of the carrier status of pigs by *not* allowing stragglers to mix with an incoming batch and for the medication of possibly infected animals.

12. Give special attention to the control of flies and rodents, which can spread disease organisms.

13. Once pigs have settled down they should be:
- mange-dressed twice, ten to fourteen days apart by enclosing pigs and soaking, using a knapsack sprayer/applicator;
- treated against worms.

14. Take care with security on the unit for:
- transport;
- birds and other pests;
- human contact.

Besides reducing the risks of transmitting pneumonia and rhinitis this will also help prevent other epidemic-type conditions, such as TGE and epidemic diarrhoea.

THE FEEDING SYSTEM

Although more detail is given in Chapter 17 on growth and feeding strategies, it is necessary to make certain practical observations at this point. Firstly, that pigs from 20 to 60 kg should gain weight at the rate of over 600 g per day on average. The feed system must allow this rate of gain to take place—as a minimum aim.

Many producers, sensibly, practise hopper-feeding at this stage and rely upon the pigs to eat enough to achieve the desired growth rate. The other common feeding method is to offer feed on the floor but, at this phase of the pig's life, the disadvantages of this method

Plate 10. The build-up of dusty, damp or stale feed in ad-lib hoppers will restrict feed intake.

probably outweigh the advantages which are lower capital cost and, possibly, higher stocking density.

Hopper feeding in the growing stage allows the following advantages.

1. More feed to be available without wastage, avoiding—

● trampling and fouling;
● loss by way of dunging system.

2. Greater intake due to—

● less fouling;
● better physical form of the feed—particularly if pelleted.

3. Less competition (providing that sufficient hopper space is available), because the pigs are not fed at intervals, but have free access around the clock.

In many instances, pigmen already using a hopper feeding system complain that their pigs eat no more than when floor fed and/or still do not achieve the rates of growth suggested above. Quite often this

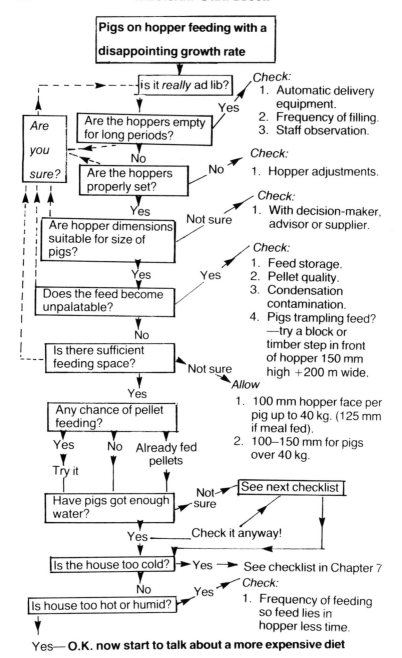

Pigs on hopper feeding with a disappointing growth rate

Is it *really* ad lib?

Check:
1. Automatic delivery equipment.
2. Frequency of filling.
3. Staff observation.

Yes

Are the hoppers empty for long periods?

Are you sure?

No

Are the hoppers properly set?

No

Check:
1. Hopper adjustments.

Yes

Are hopper dimensions suitable for size of pigs?

Not sure

Check:
1. With decision-maker, advisor or supplier.

Yes

Does the feed become unpalatable?

Yes

Check:
1. Feed storage.
2. Pellet quality.
3. Condensation contamination.
4. Pigs trampling feed? —try a block or timber step in front of hopper 150 mm high +200 m wide.

No

Is there sufficient feeding space?

Not sure

Allow
1. 100 mm hopper face per pig up to 40 kg. (125 mm if meal fed).
2. 100–150 mm for pigs over 40 kg.

Yes

Any chance of pellet feeding?

Yes   No   Already fed pellets

Try it

Have pigs got enough water?

Not sure → See next checklist

Yes —— Check it anyway!

Is the house too cold? → Yes → See checklist in Chapter 7

No

Is house too hot or humid? → Yes

Check:
1. Frequency of feeding so feed lies in hopper less time.

Yes— **O.K. now start to talk about a more expensive diet**

can be attributed to a series of errors in assuming that hopper feeding means 'ad lib' whereas examination of the checklist on page 188 may reveal that there is a 'hidden' restriction taking place, much of which may be within the pigman's scope to improve. It is especially important to remember that if the pig has to work hard for its food and water, it gives up and 'makes do' with a lower level of intake. Furthermore, badly ground pellets are less palatable to pigs than feed in a meal form.

Whilst the number of pigs per pen is dictated by others it is a fact that groups of more than 15 pigs per pen reduce average daily feed intake and, therefore, growth. The pigman must be aware of the real dangers of exceeding the theoretical capacity of pen. Overstocking may help to maintain lying area temperatures during winter but that will be at the expense of:

● feed intake due to intense competition;
● good ventilation and the removal of humidity, which further depress feed intake and pre-dispose pigs to disease.

Overstocking of pens may have little effect on the aggressive animal except that, by consuming more feed at the expense of his timid pen mate, he may become too fat, whilst the smaller pigs grow very poorly and cause the average performance to be depressed.

Pigmen should also be conscious of another important factor involving ad lib feeding. The early tell-tale sign of an unwell pig is that such animals do not eat rapidly, so hopper feeding makes it much more important for the operator to rouse and inspect every pig at least once per day in such systems.

A key element in realising feed scales during this particular phase of life, when rationing is not recommended, is that pigs can limit their own intake. A major influence upon appetite can come from the availability of water. The quantity of water required by pigs at various stages is still open to conjecture, but studies suggest that two associated points must be taken into consideration:

1. Pigs concentrate their water intake to daylight hours so it is unwise to assume that a full 24-hour access to water will be used.

2. There is a limit to the time pigs will wait for water or spend 'working' or operating drinkers, thus it is reasonable to assume that if any elements in the supply of water are at all restrictive the pig will simply drink less. Having drunk less it will eat less. Having eaten less it will gain less weight. In extreme cases of deprivation, pigs take on a drunken stature due to concentration of undiluted body salts ('salt poisoning').

Common water supply faults

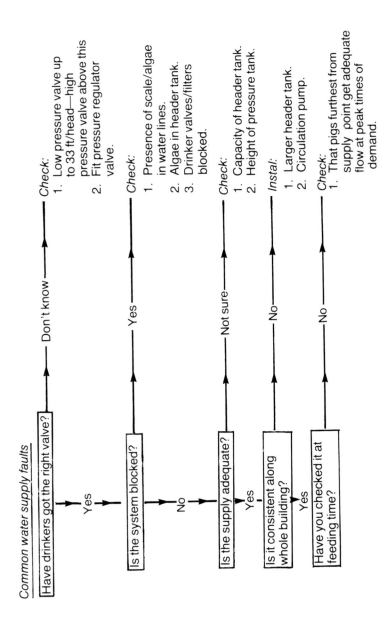

Have drinkers got the right valve?

— Don't know →

Check:
1. Low pressure valve up to 33 ft/head—high pressure valve above this
2. Fit pressure regulator valve.

↓ Yes →

Is the system blocked?

— Yes →

Check:
1. Presence of scale/algae in water lines.
2. Algae in header tank.
3. Drinker valves/filters blocked.

↓ No →

Is the supply adequate?

— Not sure →

Check:
1. Capacity of header tank.
2. Height of pressure tank.

↓ Yes

Is it consistent along whole building?

— No →

Instal:
1. Larger header tank.
2. Circulation pump.

↓ Yes

Have you checked it at feeding time?

— No →

Check:
1. That pigs furthest from supply point get adequate flow at peak times of demand.

How much water does a pig require for full growth? Estimates show that it may be in the following ranges and experience suggests that the preferred drinker types by the various classes of pig are as shown below:

Pigs' weight range	Water requirements per day on dry feeding (litres/pig)	Recommended type of drinker	Pigs per drinker
Pigs up to 15 kg	1.2	Nipple or bowl	8 10
Growers 15–30 kg	2.25	Bite	10
Young pigs 30–65 kg	5	Bite	10
Finishers over 65 kg	6	Bite	10

It is necessary to stress that provision of the recommended number and style of drinker can still fail to make available to the pig adequate water for full growth. It is this false assumption which frequently deceives the pigman into believing that he cannot and need not consider water any further than providing an automatic valve in a pen. There are still many possible reasons why there is not an adequate supply.

Current estimates are that a pig will normally only drink for up to about twenty minutes during a twelve- to fourteen-hour span each day. So in that time the pigs must be able to consume the quantities shown in column 2 of the above table. As with eating, the quantity consumed in a given time increases with the size of the pig, so requirements for drinker points may reduce as the pigs grow due to their ability to drink faster, providing there is sufficient water pressure.

The great importance of rate of flow can be appreciated from the following example:

● *Pigs weigh 30 kg and have a water need of 2.25 litres per day.* Measured rate of flow at drinker point with no spillage is 110 ml/minute (this can be gathered in a simple calibrated container such as a plastic measuring jug).
● *This means that:* to get its water requirement each pig would need to spend twenty minutes per day at the drinker to obtain its ration, presuming that it does not spill any while drinking and that it can work the drinker to full capacity.
● *But a 50 kg pig needs 4.25 litres per day.*

In the above system, this pig would have to operate the drinker for almost forty minutes a day, so is unlikely to drink sufficient to maintain full appetite and growth.

Another water supply consideration is to ensure that nipple and bite types are set at the correct height. Ideally, pigs should stretch upwards to operate nipple and bite types and so the tips of these drinkers should be sited about 100–150 mm above the back of the average-sized pig in the pen.

OTHER CONSTRAINTS UPON FULL GROWER PIG PERFORMANCE

Thus, having set out to achieve rapid growth during the grower phase and having attempted to remove husbandry obstacles from the pathway, it is now necessary to consider those other occasional hindrances which occur along the route to an early and profitable slaughter stage.

Prolapse
This condition normally occurs in pigs which are undergoing a rapid period of growth. The pigman can help to reduce the incidence of this condition by:

1. Taking care when changing the diet or feed system used.
2. Preventing a 'check' to growth, because it is pigs suddenly growing very rapidly after a stressful period which appear most susceptible.
3. Helping to control pneumonia, because excessive coughing can pre-dispose to prolapse.
4. Preventing constipation by poor access to water or dietary influences, fibre levels in particular.
5. Ensuring good pen hygiene to reduce effects of parasites, bacteria and need for some drugs, because these can cause inflammation of the rectum (colitis).

'Bent Backs'
One of the few problems associated with very rapid growth of very lean and muscled pigs. Pigs develop abnormal swelling of muscles along, usually, one side of the back during the period of life when growing rapidly (30 kg and over). The pigman can help to reduce the incidence of the problem by:

1. Removing affected pigs from the pen and marketing quickly.
2. Monitoring parentage if the herd recording scheme allows.

However, this condition is only likely to occur where pigs have a high lean capacity.

Tail Biting, Ear Sucking and Fighting

Although pigs may fight at any stage when mixed, there are occasions when spontaneous aggression breaks out in a pen. There is no doubt that some strains of pigs are more prone to fighting than others, and the pigman can only influence this factor if the recording system allows him to collate the parentage of those suspected of being the 'trouble makers'. Ear sucking is sometimes associated with pigs which are suffering from 'greasy pig' disease where the moist areas of the skin are thought to be more attractive to their pen mates.

All these problems are likely to be very complex in origin. Certainly tail docking (see Chapter 14), helps to minimise the likelihood of tail biting occurring with its attendant losses due to abscess formation and carcase condemnation. However, it will not remove the root cause, and when such aggressive occurrences arise the pigman should consider the following factors and plan some appropriate action:

- Are the pens overstocked?
- Is there sufficient feeding and drinking opportunity?
- Is there a very wide weight range within the group?
- Is the diet causing gut irritation? Not only can fineness of grinding in dry feed affect this but some raw materials, particularly by-products, can have a marked effect.
- Are the pigs uncomfortable due to poor temperature control?
- Could bedding or some 'toy' be used to overcome boredom?
- Could sexes be separated?
- Would spraying the pigs with aromatic oils or mange wash help?

Remember

That the period between the immediate post-weaning phase and just before sale is probably the most neglected period of the pig's life, *yet* if good standards are not achieved at this stage, compensatory growth may occur, but it will *not* be possible to compensate for the lost lean meat potential. The pigman should use all the checklists to eliminate any self-imposed restrictions to full growth and should not be 'feed-scale orientated' but rather 'growth-rate orientated'.

SECTION VIII

Chapters 17–18

Finishing Pig Management

Including checklists for:

Influence of stocking rate on growth
Standards for growth rate at various stages
Setting the feed scale
Appetite effects
Feeding by-products
Dirty pen floors, gut ulcers, adenomatosis, colitis
Poor gradings
Organising despatch
Re-matching and mixing of finishing pigs

Measuring growth rate
Pig identification methods
Setting of growth targets
Improving pig throughput
Improving gradings
Performance guidelines.

Chapter 17

Regardless of the weight at which pigs are to be marketed, it is essential that they reach that weight in the shortest possible time consistent with an acceptable level of carcase fatness. The trouble is that an unrealistic view of 'acceptable carcase fatness' is frequently taken. The financial reward for an extra 20 per cent of pigs in top trade almost always yields a very poor cash response if it is achieved at the expense of more than just a few days of growth.

What has been poorly recognised by the majority of pig producers is that significant changes in the potential fatness of pigs have occurred since the early 1960s. This means that the pigs are now later maturing or reach a more advanced weight-for-age before becoming as fat as their predecessors. In addition, there has been an extension of the number of entire boars being marketed for meat, which has had a further influence upon lower fat levels.

Thus the previous obsession with restriction of growth to achieve preconceived standards of grading excellence has been outdated by the modern pig. In addition there has been a failure to recognise the effects of an extended growth period on feed efficiency. While excessive feed can give rise to excessive fat, it must also be remembered that every day that a pig is on the farm, it needs to consume a 'maintenance' requirement which has to be satisfied before the pig gains any weight whatsoever—see fig. 15.

In this example, the difference between the columns—which

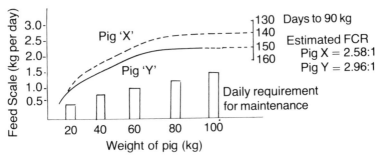

Fig. 15. Feed available for growth

represent the body maintenance requirements—and the curved lines shows the amount of feed available for growth. Thus, although Pig 'X' is given more feed per day, it lives for thirteen days *less* than Pig 'Y'. Because all of this saving occurs late in the pig's life, it eliminates maintenance at the *higher* rate, so the calculated feed efficiency difference in this example favour Pig 'X' by 0.4:1. Whilst such a difference may be difficult to measure on the farm, this is a realistic illustration of the probable trend.

A final reason for adopting a policy orientated towards growth rate concerns the use of the facilities provided. On the majority of units pen space is provided for a given throughput of pigs based upon arbitrary stocking levels. With the exception that *very* low stocking rates may adversely affect temperature control in cold weather, it is safe to expect that lower stocking rates will give quicker growth due to the beneficial effects of reduced competitition. Thus, it is possible to plan for a growth bonus:

1. Grow pigs faster to make lower stocking rates possible.
2. This leads to *bonus growth* due to lower stocking rates.

As mentioned previously, many of the pigman's problems have their initial roots in too heavy stocking rates. This is the one way in which the pigman can reduce those difficulties *without* increasing the overhead cost of buildings which is spread over all pigs sold.

Conversely, if throughput is not as fast as predicted, an ever-expanding backlog of pigs occurs and this all too common sequence goes like this:

1. Pigs grow slower so there are more pigs on the farm at any time.
2. This leads to overstocking of pens so pigs grow slower due to:
● competition;
● possible aggravation of health problems;
● vices.
3. This means that pigs are either:
● moved into houses smaller than they should be which compounds the slow growth effect still further; *or*
● sold at lighter weights which means that the overhead costs of the unit have to be spread over a lower total weight of pigs sold.

Even apart from all the above arguments in favour of faster growth, there is perhaps the best reason of all for close attention to good growth rate from the pigman's standpoint. Growth rate is the *only* measurement that the pigman can make on a day-to-day basis and use to influence pigs going through the system *now*. Feed conversion ratio is vital to the owner, but the pigman can only measure it retrospectively and, even then, with great difficulty. As there is such a close connection between good growth and low feed conversion, it is reasonably sure for the pigman to assume that:

If pigs are growing well, particularly early in life, they will be converting their feed well.

The strategy by which a pigman may measure growth and use the information is explained in Chapter 18.

How Fast Should the Pigs Grow?

It is impossible to provide yardsticks for growth rate which will be acceptable under all circumstances, because type of pig and diet used can quickly produce pigs of quite unacceptable carcase quality.

In general terms, it pays to aim to get around 70 per cent of pigs in the top payment grades if pigs are sold on a deadweight basis with a differential price for different fat thicknesses. Because this level of carcase grading can be achieved by so many routes, it is possible to use the suggested levels below as a guideline only. More accurate data for your own herd can be achieved by following the suggestions in Chapter 18.

The first aim should be to achieve the growth rates shown *at the very least* and then to adjust management and feeding to suit the desired level of fatness.

Age of pigs (weeks)	Suggested growth pattern (g/day)
3– 6	380– 420
6–10	460– 500
10–14	560– 680
14–18	740– 850
18–22	900–1,000
22–26	870– 900

How Much Feed Should the Pigs be Given?

To achieve the levels of growth rate suggested the answer is: *However much it takes to achieve those levels of growth which give the desired carcase fatness.*

The time is long overdue when pigs should cease to have their performance constrained by feed scales which have not been adjusted for modern circumstances.

Thus, the recommendations are:

- do *not* feed pigs according to their weight because they may not weigh enough relative to their age;
- rather feed according to the age;
- *but* preferably feed according to their weight-for-age (see Chapter 18).

WHAT CAN BE DONE IF PIGS WILL NOT EAT ENOUGH?

Once pigs reach about four months of age, feed intake, given satisfactory circumstances, ceases to be so much of a problem. However, at any stage during the growing phase it may be necessary to assess the influences upon the pig's appetite. Indeed, on some units it is still a problem to encourage pigs to eat enough so that the pigman feels frustration at his employer's urges to increase pig thoughput.

After all, a pigman may be asked to speed growth rate by 5 per cent and, in most cases where feed wastage can be controlled, this can be achieved by increasing feed intake by a similar percentage. Where should the pigman look if the pigs refuse to eat the extra 5 per cent?

It has to be said that there may well be a breed or type effect on appetite so that some pigs have a lower inherited appetite capacity as a result of their genetic make-up. The pigman cannot expect, or be expected, to overcome such a constraint in the short term, but it will make consideration of other points indicated more important.

Feeding system is something else which is beyond the scope of the pigman to influence. There seems little doubt that as feed levels are increased closer to the limits of pigs' appetite, a *descending* order for influencing appetite would be expected as follows:

- three times a day trough feeding;*
- twice a day trough feeding or ad lib;
- three times a day floor feeding;*
- twice a day floor feeding or once a day trough feeding;
- once a day floor feeding.

The reason for the lower intake from floor feeding is likely to be the effect of contamination, which has an increasingly depressing effect on intake the closer the feed scale is to appetite. So, although

* This extra feed will only make a difference of 2–5 per cent if there are at least six hours between timings of the feed. Simply inserting one midday feed into daily routine has no, or little, effect on intake.

the pigman cannot alter the system, he can to some extent influence intake by frequency of feeding.

In addition, the pigman will not be able to control feed presentation, but any dusty pellets or poorly ground meal should immediately be brought to the attention of whoever makes the

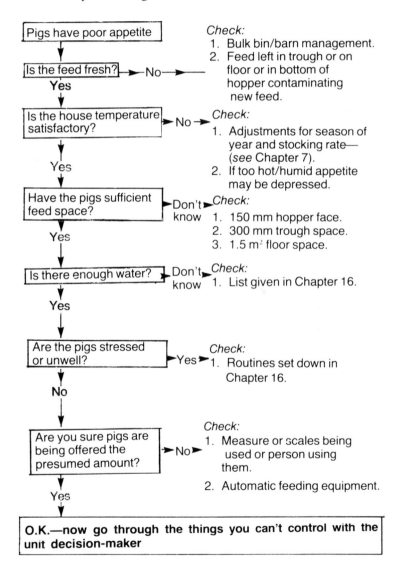

Pigs have poor appetite

Check:
1. Bulk bin/barn management.
2. Feed left in trough or on floor or in bottom of hopper contaminating new feed.

Is the feed fresh? →No→

Yes

Is the house temperature satisfactory? →No→

Check:
1. Adjustments for season of year and stocking rate— (*see* Chapter 7).
2. If too hot/humid appetite may be depressed.

Yes

Have the pigs sufficient feed space? →Don't know→

Check:
1. 150 mm hopper face.
2. 300 mm trough space.
3. 1.5 m² floor space.

Yes

Is there enough water? →Don't know→

Check:
1. List given in Chapter 16.

Yes

Are the pigs stressed or unwell? →Yes→

Check:
1. Routines set down in Chapter 16.

No

Are you sure pigs are being offered the presumed amount? →No→

Check:
1. Measure or scales being used or person using them.
2. Automatic feeding equipment.

Yes

O.K.—now go through the things you can't control with the unit decision-maker decisions, as should any sudden rejection of feed, which suggests contamination or the pigs' response to a change in raw materials.

The pigman's own checklist of items which he can influence and which, in turn, affect the pigs' appetite is shown on page 201.

Why not just hopper feed? Where pigs of high genetic merit are being sold uncastrated in surroundings which allow rapid early life growth, it may be feasible to ad lib feed pigs through to slaughter and still achieve acceptable carcase quality. However, the pigman should be aware of:

- increased fouling of solid floor areas when pigs exceed 50 kg in weight;
- the need to check hoppers carefully for stale feed;
- possible feed wastage if hopper design or setting are poor;
- the need to inspect pigs closely at least once per day in order to identify any early signs of poor thrift;
- the need to manipulate slaughter weights to avoid gilts becoming too fat;
- the possible need to control the upper limit of daily feed intake to achieve desired levels of fatness;
- the need to prevent bullying and over-stocking, both of which can lead to a wide range of growth rates within groups of pigs.

Results show that whilst increased feed intake and growth rate may occur as a result of hopper feeding, feed conversion ratio may be poorer pointing to the importance to minimise spoilage and waste. The shape of the pen may influence intake and wastage and inasmuch as a narrow pen will make it difficult for pigs to feed at a position of 90° to the feed face and so waste and competition may increase.

Special Problems Associated with By-product Feeding

Most by-products are in bulky or liquid form, so they are normally associated with trough and pipeline feeding systems. Most by-product feeding systems suffer from a greater unpredictability in raw material analysis and this may make the setting of feed scales more tricky. These systems can, however, give very cost-effective methods of production and it is worth drawing attention to some of the major points which a pigman should concern himself with when using the most common by-products—those from dairies. Although other diverse materials are used, the main considerations recommended here can be applied in principle to most materials used.

1. Always take a sample of each delivery and label it clearly for analysis and advice on accurate inclusion rates.

2. Mix as accurately as the system allows.

3. Always treat the by-product as a valuable raw material, not a *waste* product.

4. Ensure that storage tanks are cleaned.

5. Always feed the material in the same state of freshness. (1–1½ litres formalin in 1,000 litres of skim or whey delays scouring.)

6. Monitor carefully feed scales because over-feeding can lead to 'whey bloat' and 'twisted gut' conditions.

7. Never store by-products in galvanised containers.

8. Always make water available even when feeding high levels of milk by-products because whey, in particular, may have high salt content.

9. Be prepared for frosty weather by lagging pipes and valves to prevent freezing and depressed pig appetite from very cold liquids.

OTHER FEEDING HERD PROBLEMS

Most of the problems discussed in earlier chapters will also occur during the final stages of the pig's life. The problems described here, similarly, may arise at earlier stages of the pig's life.

Dirty Pen Floors
This is one of the most frustrating, unpleasant and time-consuming problems which face a pigman. There is no one cause or cure, just like tail-biting, and the pigman has to attempt systematically to check off the details in the list below, keeping a careful check on his modifications and their effects.

1. Check the air movement pattern in the house; can adjustments be made to make the designated lying area more attractive to the pigs?

2. Is there a difference in air movement at night compared to day time?

3. Is the choice of lying zone adequately demarcated?

4. If all-in/all-out is practised, is the ventilation system sophisticated enough to cope with this?

5. Is the stocking rate too high or low?

6. Is the dunging area too small for the number or size of pigs housed?

7. Are the drinkers leaking?

8. Can shavings or chopped straw be given to encourage correct lying pattern?

9. Can a change of floor feeding be used for problem pens?
10. Can the ventilation controller be adjusted so that differential temperatures and air movement patterns are achieved between day and night?
11. Can an adjustment to feed levels be made to influence pig behaviour?

Gut Ulcers
Normally associated with low feed intakes for prolonged periods. The pigman should be aware that:

● More concentrated diets, particularly in meal form, may pre- dispose pigs to this condition, so competition for feed must not be allowed to cause very low feed intakes and you should be on the lookout for 'pale' pigs.

Adenomatosis (P.I.A.)
This is a gut condition which leads to thickening of the gut wall ('hosepipe gut'), poor feed absorption and eventually death. The herd with the, otherwise, healthiest pigs may even be at greater risk due to the higher incidence amongst minimal disease popu- lations. As soon as one case has been diagnosed by a veterinarian, the pigman should be on the alert:

1. To spot and inject affected pigs with the prescribed drug quickly.
2. For any otherwise healthy pig losing condition very rapidly.
3. For pigs showing signs of dark, reddish-brown diarrhoea.

The need to medicate pigs routinely after a stressful move or transfer often follows diagnosis. Alternatively, it may simply be necessary to inject individuals affected or, on occasions, to medicate the whole growing/finishing herd for a short period via the feed.

Colitis
This gut condition is, for most purposes, similar to P.I.A. with the exception that the scour observed is not normally discoloured. The treatment is the same as for P.I.A. and prevention of both these conditions involves:

1. Improved attention to pen hygiene.
2. Reduction of stress caused by overstocking and poor environment control.
3. Greater care at movement.
4. Rapid diagnosis and treatment by the operator.

Poor Gradings

Poor Gradings

Chapter 18 details the feeding regime manipulation which must be contemplated in ensuring that there is a balance between feed intake, lifetime growth pattern and tolerable carcase composition. In addition to feeding practice there are a number of other features which an operator should consider. These are:

1. Are sexes being treated differently – boars are less fat, at the same weight, than gilts which are leaner than castrates?
2. Can feed scale be controlled in final period of life?
3. Is a differential scale weight being operated to ensure that the potentially fatter pigs (castrates/gilts) are being marketed at a lighter weight than the leaner ones? As an example, gilts marketed between 82-87kg with boars sold at 92-97kg would still give an average slaughter weight of 90kg and would help to balance the inherent carcase features dictated by sex. Thus, optimum *average* contract weight may be balanced by optimum fatness levels.
4. Can stocking rates be adjusted so that the more dominant pigs are not feeding at the expense of the timid pigs?

Optimising the sales outlet has a crucial impact upon profit. Pigmen should not become totally obsessed with individual components of performance such as backfat level, F.C.R. or even speed of growth but should be prepared to adjust husbandry practices which allow the correct balance of physical performance to gain optimum profit from the chosen market outlet.

Organising Despatch Arrangements

Whilst the importance of achieving farrowing targets was stressed at the 'input' end of a breeding and feeding enterprise, it is equally important to:

● schedule growth;
● organise sales in the prescribed weight range.

If this is not achieved, the inevitable bottleneck of facilities will arise which produces a self-generating violent downward spiral of lowered efficiency.

Thus, the pigman needs to arrange for growth rates to balance the predicted pig output from the unit and to monitor closely that pigs leave the unit in the number in which they enter with an allowance of no more than 2 per cent losses en route.

Within reason, the closer pigs are to the upper limit of the allowable weight range, the greater the profit potential, providing that this does not give rise to the 'dreaded' overstocking fault. Thus, the pigman should watch closely for a dropping-off in sales as an early indication that pigs are not growing as fast as they are scheduled to do.

Despatch organisation is the only way that the pigman can regulate his hygiene and pig movement flows, and it is vital that potential bottlenecks are seen in advance so that solutions can be sought (like selling some pigs younger to allow normal routines to be followed). Alternatively, any temporary surplus penning may be useful if maintenance can be planned to take place during a short lull in pig throughput. In this way, even a shortage in sales can be put to full and effective use.

In order to capitalise fully upon the weight bands dictated by a chosen outlet it may be convenient to withdraw lighter pigs from a number of pens for sale in a subsequent week. This poses a pigman with a problem of how to use pen space effectively yet avoid potential losses caused by mixing of large pigs. Clearly, it is preferable to avoid mixing wherever possible but optimising a sales outlet and re-batching to avoid overstocking of pens are two instances where a sound technique of mixing may be useful.

Pigs near the end of the finishing period which fight will quite possibly be heavily penalised at the abattoir because of bruising and skin blemishes. Thus, if mixing is attempted and is followed by price penalisation at the slaughter point, it may have to be avoided *or* re-batching practised at least two weeks prior to likely sale. Some producers refuse to contemplate re-batching and prefer to sell the slower growing pigs to a separate outlet but this may be less beneficial from a financial viewpoint.

The following considerations will help to ensure trouble-free mixing of even quite large finishing pigs:

1. If possible mix pigs from more than two groups.
2. Try to avoid extreme weight ranges – maximum 10kg.
3. Move all pigs to a different pen if possible.
4. Avoid mixing during very hot weather and delay to end of daylight hours if possible.
5. Consider enclosing pigs in a narrow pen initially (for the first 2 hours) to minimise opportunities for aggression, (maximum 2m wide if available).
6. Provide bedding if possible.
7. Most important – spray pigs comprehensively, paying particular

attention to the head and tail areas, with a proprietary aromatic wash to deter fighting.

8. On releasing pigs to main pen offer feed and water immediately to distract them from territorial behavioural aggression.

9. On releasing to main pen allow more generous space tolerances than normal.

Where this technique is followed experience shows check-free growth and negligible carcase damage can be achieved.

Chapter 18

If the strategy that the growth of the pigs should not be controlled by a predetermined feed scale but by the speed of the animal's growth relative to the required throughput of the unit is accepted the pigman must establish a clear-cut husbandry regime.

As mentioned previously, it is not entirely acceptable, or appropriate, to use a feed scale which is geared to the age of the pigs alone. To adopt a non-varying feed scale geared only to the age of pigs in the hope that they will catch up is unacceptable. For true progress and the achievement of a higher level of output, it is necessary to adopt a management approach just as sophisticated and well monitored as that used in the breeding herd. For too long the finishing herd has been the neglected part of pig production. Yet it is that part of the pig enterprise where the majority of the feed is consumed. So, because feed comprises the major cost of pro-

Plate 11. Occasional check-weighing of pigs is a vital element of precise finishing herd management control.

duction, there is a greater opportunity to improve the profitability of a unit by concentrating effort in this phase of production.

The adoption of the proposed schedule—based upon achieving a given standard growth rate or weights-for-age—has its origins in the need to know as soon as possible that pigs are not achieving the expected level of performance. One of the major problems with using food conversion—important effect though feed usage has upon profit—is that it provides the pigman with only retrospective information. It is also difficult to measure, so the pigman—even if an accurate feed conversion figure can be regularly achieved—is only being told that things have *not* gone well in the *past*. It also means that a further number of pigs still on the unit have gone a substantial distance through their growth span and it will be substantially too late to influence them.

Further, the use of carcase gradings as the sole arbiter of finishing pig performance can be a poor indicator of profitability because the reward for good gradings may not justify the extension of the growth period by very many days. In addition, it provides, again, only retrospective information.

Measuring Growth Rate

The pigman must be able to identify the age of the pigs. Because a degree of fostering may, and a mixing of litters will certainly, have taken place, the precise date of birth may not be strictly essential except to a specialist breeder.

Most commercial producers would find it adequate to be able to age a pig in weeks rather than days, and rely upon a typical spread of farrowings to average out the age of the pigs born during the week.

Next each week will be given a number. That number is designated to all the pigs born in that week. Commonly pigs born in the first week of January will be given the number '1', and so on throughout the calendar year.

Providing every other week is numbered, the pigman then only has to subtract the pig's ear number from the week number to determine the pig's age. For example, in week 20 a pig numbered 9 would be $(20 - 9) = 11$ weeks of age. This allows the pigman to check quickly the age of any pig on the farm to assess its progress.

Identification

The safest and most convenient form of identification is one which:

● is given to the pig close to birth before fostering (if accurate parentage records are required);

● is carried by the pig as an indelible mark, i.e. as an ear notch, or
tattoo.

Ear Notching
Ears can be notched at birth to a given code (see fig. 16 for code and
use). This is easily read and there is little chance of it becoming
illegible over time. It is also quick and easy to undertake. The
pigman may refine the system by using:

● punched holes to code particular parentage by allocating each
boar a given punch combination;
● alternatively, for occasional checking of new boars, new treatments etc. the basic notching system may be supplemented by
ear tags.

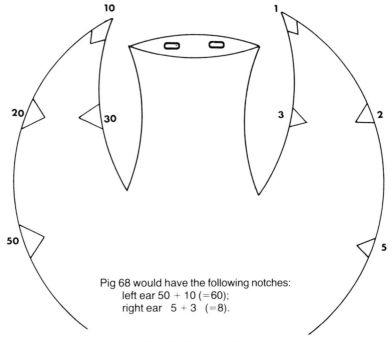

Pig 68 would have the following notches:
left ear 50 + 10 (=60);
right ear   5 + 3   (=8).

Fig. 16. Ear notching

Ear tattooing
If carried out properly, ear tattooing is a more flexible system than
notching because, with only two digits required for the weekly age
code, a wide combination of letters and numerals can be used to

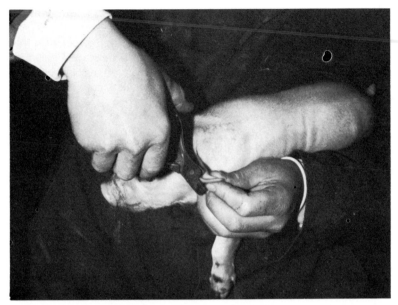

Plate 12. Ear-notching with a code to allow weekly checks for the age of pigs will permit a rapid assessment of weight for age against a pre-set target.

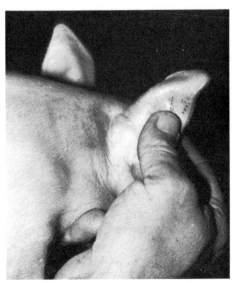

Plate 13. Tattooing each pig born with a number designated for that week allows a quick and easy check on age.

allow the monitoring of the precise coded details of sire, dam and even housing systems used earlier in the pig's life.

However, ear tattooing is more time-consuming to perform than notching, and may be less easily read at a distance, or if pigs have dirty or coloured ears. It has a further disadvantage of being less easy to carry out very early in the pig's life.

Ear tagging
Once again, an infinite combination of colours and marking can be achieved with tags. These advantages are offset by the need to use a very small tag for new-born pigs which will be difficult to read later in life, and the risk of animals losing their tags.

Body tattooing
A form of slap marker can be used to create a permanent mark—usually in the softer, hairless, area on the pig's neck just behind the ears.

The major drawback of this system is the need to use it in conjunction with some other form of identification for the period earlier in life before the body tattoo can be used which may be around twelve weeks of age.

Pen card system
As an alternative to actually marking the pig in one of the above forms, each litter and then pen may be given a card which follows the pigs around the unit.

This system has the advantage of allowing details of the management, feeding and even parentage of the pigs to be noted on the same card.

However, the pen card approach becomes less easy to manage where pens are split and re-mixed, and there is always the risk of accidental damage to, or loss of, a pen card, destroying the information on a group of pigs.

SETTING THE GROWTH TARGETS

Once the age of pigs is organised by one of the identification methods outlined above, the pigman next needs to establish the growth rates being achieved by the herd. This will mean weighing at given stages throughout the life span of the pigs.

Initially, the pigman probably has two standards of weight for age. He probably knows, with reasonable accuracy, the weaning age and weight and probably also the average age at slaughter.

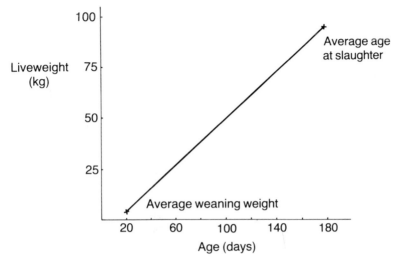

Fig. 17. Growth pattern with only two known weights for age

Given this information, he can presume only that the pigs grow on a straight line pattern between these two points, although, in actual fact, this will never be the case. Thus, it is important to progress beyond this point which is illustrated in fig. 17.

If this growth rate is below expectation, the pigman may immediately assume that he needs to take some action. The easiest move is to increase feed scale in that period following ad lib, or appetite feeding—in other words, during the final stages of growth.

Such an action is the usual response to a disappointing level of pig throughput, *but* it will probably yield faster growth at the expense of a marked decline in grading standards. This is because, as explained in Chapter 16:

● efforts to gain weight rapidly late in life following modest growth up to 60 kg will produce fat in an adverse ratio to lean;
● it will probably also have a disappointing influence on herd feed conversion.

What the pigman requires to influence finishing herd performance is more information about his pigs. Ideally, a growth curve for the unit will be obtained by weighing pigs across the whole finishing period and plotting these on a graph.

The thought of weighing large numbers of pigs in an arrangement not designed for high-speed weighing is a daunting prospect for the

busy pigman. There are, however, several options for him to consider in establishing such a log of information.

1. To weigh every pig, noting its age at the same time and
transferring this information to the graph (see fig. 18).

This provides the most accurate indication of the current rate of
herd growth.

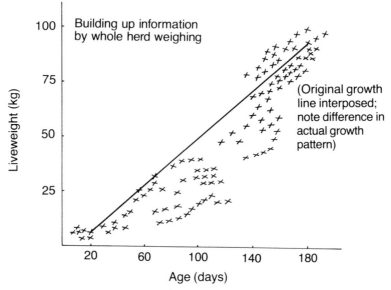

Fig. 18. Building up information by whole-herd weighing

2. To commence the check weighing of pigs as they are moved
from one house to the next. This has the advantage of giving several
valuable pieces of growth information without creating any
additional pig movement or disturbance. As it is usual for pigs to be
moved at least twice in their finishing life, this provides extra
accurate points to add to the actual growth curve which may typically be augmented by last-month weighings when pigs are being
screened for slaughter. See fig. 19.

3. Alternatively, certain groups of pigs can be selected as
'markers' and used to obtain basic information against which subsequent batches may be compared. If this is started at a given point
so that management, feed, health effects can be attributed to the
performance levels achieved this technique will take longer to build

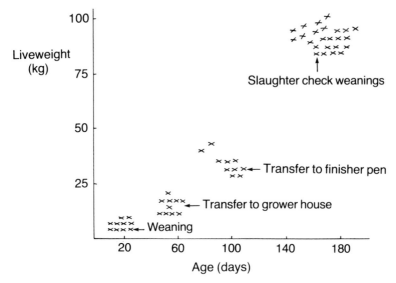

Fig. 19. Information gathered at house transfer

up. As a guide 20 per cent of each group of pigs moved yield accurate and useful data.

4. Alternatively, the 'straight line' graph may be drawn or the growth rate figures suggested in Chapter 16 used, or even those obtained from other producers, and then the pigman can begin to use the information he is about to gather against the average or target levels. Finally, the pigman may devise a target growth curve from the rate of growth that he needs the pigs to achieve to avoid over-stocking the farm.

USING THE GROWTH RATE INFORMATION

Given the base by which the growth rate of other pigs can be reasonably compared, the pigman can now build up a log of data, probably every time the pigs are moved, by weighing at least 20 per cent of the animals.

Thus, within a brief period of time, the pigman may learn a great deal about his pigs.

If some common and typical performance trends are examined, clues to the use of the weight-for-age technique can be given.

Example 1

In fig. 20 we see a typical set of farm circumstances. The pigs are falling well behind the desired throughput for the unit, which may well be leading to overcrowding problems making it very difficult for the pigman to operate at the stocking densities that he *knows* are right for the buildings provided. Notice, however, that the average P_2 level is low, suggesting that very good gradings would be achieved.

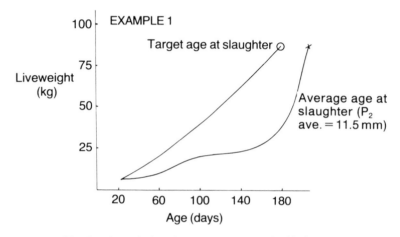

Fig. 20. *Actual slaughter age compared with the target*

In this case he would consider this checklist:

1. Is the feed scale correct—particularly earlier in life?
2. Young pigs may be kept on ad lib feed longer and on creep/grower rations longer in order to achieve better early-life growth.
3. If the feed scale seems satisfactory, is the feed actually being fed—check scales, automatic equipment, recording and wastage.
4. Is the feed scale suited to the environment of the house?
5. Are the pigs suffering from competition for space, feed or water?
6. Observe pigs closely for any signs of ill health.
7. Having checked the above *first*—consider diet quality and feed additives.

Example 2

Again, fig. 21 shows a common set of circumstances. In this case, almost the opposite of the first example, the pigs are reaching

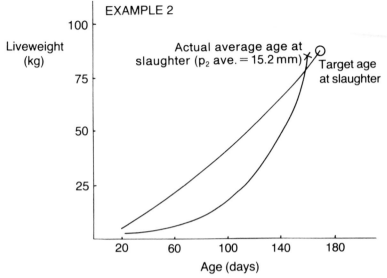

Fig. 21. Grading and growth

slaughter rapidly but, because of the growth pattern achieved, the pigs are grading badly. Note the average P_2 figure.

With this growth information the pigman would consider this checklist:

1. The feed scale—in this case the pigs may be overfed, particularly in the later stages of life.

2. Concentrate on improving weaning weights (see Chapter 14).

3. Improve weaner management—in particular give below-target pigs special management consideration (see Chapter 15).

4. Check environment in early stages of life.

5. Is the ration too dense in later stages of life?

6. Is the stock good enough? (See notes below.)

7. Is it possible to cease castration, or can pigs be penned so that differential feed scales can be applied to pigs of different sex? It is important for a pigman to remember that on the same feed scale:

● boars will grow 5–7 per cent faster than gilts;
● gilts will grow 5–7 per cent faster than castrates.

As pigs grow faster, that is, are offered more nutrients, this difference may manifest itself by:

● castrates being 5–7 per cent fatter than gilts;
● gilts being 5–7 per cent fatter than boars.

It is a sound practice to separate the sexes at the weaner or grower phase and to adjust feed scales which account for the above differences in sex. The pigman should remember that the sex difference will be exaggerated: (a) when pigs are fed on higher feed scales and better diets, (b) when pigs are finished to higher weights.

Example 3
Another valuable use of a known weight-for-age scale for a unit is that any change in production method such as diet, feed scale, housing or penning arrangements can be easily checked and accurately evaluated.

SOME PERFORMANCE GUIDELINES

Although the extent of carcase change will vary between pigs of different genetic potential, it is possible to predict with the benefit of some experience that:

1. Carcase fatness may change not just by altering the speed of growth, but by:
Changing carcase weight—if the average weight at slaughter falls by 1 kg deadweight, the average P_2 will drop by 0.2 mm.
Thus, a change in marketing policy must not be confused with management effects.

2. Carcase length may change by:

● up to 1 mm for every day that growth rate is altered;
● approximately 2.5 mm at the same weight with boars being that amount longer than gilts and gilts the same amount longer than castrates.

And finally . . .
It will not be necessary to weigh pigs as regularly once the desired standard of performance has been achieved. This may take many months of manipulation and adjustment of all those items discussed in the four chapters and it is possible that some unforeseen disruption and frustration may occur which affects pig performance and makes interpretation of the results more hazardous.

Throughout this handbook there has been little reference to genetics although source of stock is one of the items which creates most discussion in our industry. Genetics have not been discussed because the pigman cannot influence them. However, if boars and sows are retained for a long time, and the herd replacement programme is badly organised so that slaughter generation gilts have to be retained for breeding, perhaps the progeny from such parentage will not perform to expectations in the finishing house.

Thus, the pigman even influences the actual realisation of those advances made by the breeder. It is a pity if he fails to make the most of all this technology by failing *really* to apply all his skills. Now, if the reader would honestly like to make more use of his abilities—move on to the last section . . . .

SECTION IX

Chapters 19–22

You, Yourself and Others

Including checklists for:

Awareness of unit problems
Allowances for other people's opinions and preferences
Communication
Influencing others
Gathering information
The owner's problems
Finance and the pigman
Payments, rewards and incentives
Motivation
Improving understanding, knowledge, and ability and job satisfaction.

Chapter 19

I was once offered an opinion that the essential items which are needed to produce profitable pigs are 'good buildings, good feed, good pigs and good luck'. Experience has taught me that the most serious omissions in that simple summary are 'good staff and good communications'.

Good results are achieved under a range of conditions which is so wide as to make a nonsense of emphasising the importance of 'building, feed and genetics' to the exclusion of all other factors. Certainly, building design places limitations upon output as will the quality of feed and stock; however, they impose the ceiling of attainable excellence and must not be confused with that element which controls the *degree* to which these factors influence results— namely the pigman.

It would be improper to elevate the importance of the pigman's role to the point where the individual assumes that his or her position is more important than that of the owner or manager. However, a major cause of the variation in performance between farms is often the failure of the owner or manager to appreciate the impact which those undertaking the daily tasks on a unit have on output and returns.

Communication and motivation come easiest to the self-employed pigman, the owner–operator. The need to be self-critical, to seek out ways of improving performance are illuminated; there is no need to pass on decisions to others and rewards for efforts are received in the most tangible forms.

An awareness that the same needs exist when others work with, or for, us would greatly assist in achieving high levels of output. This awareness will not solve the problems of communication, but will establish a better understanding with those who are expected to show the same commitment as the owner.

In almost all cases, there need be no doubt that the pigman likes, or is fascinated by, the pig itself, and this interest in the animal can be taken for granted. Pigmen hate to see pigs suffer or die. However, although we need not be concerned with negligence reducing the level of achievement, on the other hand it is important to ensure that everyone is correctly motivated by the economic requirements of the business.

The properly motivated and informed pigman is perfectly capable of masking the shortcomings imposed upon results by less-than-perfect buildings, feed or stock. He may never optimise the total physiological capacity of his pigs, but he can do so within the constraints placed upon him by the conditions under which he is working.

To do this, however, the employed pigman must become aware of the vital part he plays in the success, or otherwise, of a unit. In addition, it is up to those who employ to ensure that clear-cut standards of achievement are understood by everyone and the means of achieving them are accepted by all involved. In other words, the pigman is a vital cog, not the big wheel, and as such needs to feel part of the 'driving force' of a smooth-running machine.

The human relationships within any group of people are very complex and we all react in an individual way to any problem. Ideally, all those in a group will act in a restrained, frank, yet logical manner in dealing with a given set of circumstances. The differences of each of our characters make this unlikely so, at best, we should all be prepared to make allowances for each other's attitudes, reactions and opinions.

The pigman himself must be prepared to think in this way. It is of no value to heap blame for poor unit performance or poor staff relationships upon those more senior in the management chain alone. There are bound to be weaknesses in any partnership, but they will never come all from one party.

If such an attitude of tolerance can be made to prevail, the very essence for a close and successful working relationship will be present. Acknowledging the differing but important interrelating parts that the managing and the managed play at all levels is necessary to the achievement of good results.

It will be necessary to attempt to influence the opinion of other members of the team from time to time. A new idea might well be useful, but it is best implemented by the general agreement of every-one concerned. This cannot be achieved by coercion, or the belittling of those responsible for a previous decision or practice. Even the least experienced trainee pigman can think. He should be encouraged to do so without being led to expect promotion or some materialistic reward because of his contribution to the success of a series of tasks.

Everyone requires his or her emotional needs to be satisfied. While most of us prefer an orderly, organised routine for our working life, a degree of self-expression and fulfilment from daily work leads to greater satisfaction. Thus, because *we* like to be consulted when any changes are necessary, so we should adopt the same approach to those junior to ourselves. It must always be remembered that some people have less enthusiasm than others and are less ambitious, but they can be solid, dependable team members and, as such, should be encouraged to express their views on any matter involving change.

A further essential attitude of mind in the pigman is that he must accurately represent what detailed points of husbandry are actually practised on the unit. All too often there is a gradual change from what is believed by management to be the routine on a unit. It is no good concealing such a change, otherwise a whole series of judgements and alterations might occur, based upon a false set of circumstances. It is quite common to be confronted with a description of a feed scale which, when totalled, bears no conceivable relationship to the overall quantities of feed purchased. It is not culpable dishonesty at work, but a failure on the part of the operator to notify others of amendments to the assumed practice, or even a fear of having to admit that some modification to routine has been made without authorisation. An atmosphere in which changes occur naturally and are understood by everyone involved must be the aim at all levels in the production unit. This is why there are references in certain sections of this book to a pigman needing to make sure that certain items have been properly checked.

There is no need for the pigman to study the philosophers to be a successful team member, but it is unlikely that he will be a successful operator without the consent of others upon whom he relies for support or guidance. As a plain guide, it is worth remembering that problems with one hundred pigs make life difficult, but problems with one colleague may make life impossible.

Because people choose to work with stock, their basic requirements are likely to be satisfaction with their job, a desire to be accepted as a key member of the team and a fair reward for results—and probably in that order.

Thus before the senior pigman can hope that others will work diligently for him, or before the junior pigman can hope to persuade those to whom he is responsible, it is necessary:

- to be aware of the unit design weaknesses, but never to use them as an excuse for lower output;
- to understand the limitations of those with whom we work;
- to be understanding of other people's domestic problems without allowing those circumstances to be used as an excuse for unacceptable work standards;

- to ensure that every member of a unit team understands what is expected of him or her and what the consequences will be if they fail to carry out specified routines accurately.

Remember
These points do not mean that it is possible to establish a smooth working relationship in every single case. It is obviously wrong to permit unit performance and staff morale to suffer from the influence of just one individual. However, if everyone is encouraged to think positively, it is reasonable to expect that better and happier circumstances will be produced.

Chapter 20

Everyone involved in pig production enjoys becoming better informed by visiting farms, meeting those associated with pigs, picking the brains of research workers and reading about new techniques. By these means they are learning from others how to solve problems on their own units and how to improve results.

This attitude, so strong in the pig industry, implies that a degree of change will inevitably take place upon one's own unit. The pigman may have one of several possible roles to play in this. Firstly, he may be the prime mover in the chain and, therefore, will need to convince others of the merits of his case. Alternatively, he may be one of the recipients of someone else's idea and will need to accommodate it into existing practices.

To adapt to a change is almost always easier than trying to persuade others to make a similar adaptation. It is frequently made more difficult because of the complexity of human relationships outlined in Chapter 19. Quite often, too, a really sound pigman is promoted to a position of responsibility for people and then finds it difficult to mould the group of individuals into a committed, successful team.

The importance of team work has been acknowledged for centuries. The attitude within a team reflects the relationship between the individual members of the team and, obviously therefore, the leadership style used. In the situation where the pigman is also in a position of responsibility for others he can only comfortably operate within the broad constraints of his own character. For example it is of no real use someone of quiet rather introspective manner suddenly adopting an authoritative 'Do it my way, or else . . .' leadership style.

Those in a position of responsibility will find it pays to separate stockmanship and social considerations in their attitude towards leadership. For example, it is quite acceptable to be firm and curt in dealing with such matters as timekeeping or tidiness. On the other hand, the conduct of general stock tasks and unit practice will be much more successful if they are the result of the participation of everyone involved. In short, we can modify tasks which affect the pigs by discussion, but there is no excuse for continual anti-social 'community' attitudes within a team.

Very few people have all the desirable leadership qualities—even if it could be agreed what such qualities are. Clearly a pigman who has both sound knowledge and innate ability will be more readily accepted by others junior to himself. However, a sound technician may not be a natural man manager. Therefore, it is necessary to consider the way in which the abilities of others can be harnessed if successful leadership is to be practised.

The pigman is mainly concerned with solving or mitigating the effects of various 'problems'. In order that an effective change can be brought about, he should consider the following checklist.

1. How will others react?
- It is important not to attempt to impose a change on someone who has a domestic crisis or when some of the staff are on holiday or sick-leave.
- Make sure that a change in routine will not bring incompatible characters into closer contact.
- Ensure that the alterations do not cause extensive changes in working practice or responsibility. Otherwise the willingness of others to accept a new process may be masked by a dislike of the need to work unsocial hours, or the desire for greater rewards, or because they give rise to feelings of insecurity over employment or seniority.
- Observe others closely for early signs of dissatisfaction with existing conditions or with any suggestion of change.

2. How will the owner react?
- The pigman does not need a crash course in economics to sympathise with an employer's need to see an acceptable return on his investment, or to reduce his borrowings (see Chapter 21).
- It is important to realise that even limited facilities can be made to work better with good operational management and, if the pigman can show an improvement in output, the owner may be more willing to consider increasing his investment.

The more successful units tend to be those where the owner, or his agent, involves the production personnel in the basic discussions about required production standards and, in turn, where the pigman adopts a similar approach when establishing routines, staffing rotas or changes in these items.

This technique calls for a degree of communication and involvement which is far greater than that practised on the majority of units. All too often communication is like a series of games: 'O' Grady Says' leading to 'Blind Man's Buff', followed by 'Chinese Whispers' and, finally, 'Charades'.

How can the would-be senior pigman avoid such a sequence? How can he or she ensure that the expectations of the owner are met and how can the whole plan produce good results and give satisfaction to those who will be required to undertake the tasks? The following checklist should be used.

1. What is happening now?

● Before any change is considered, a basis for establishing the need for a change and measuring subsequent effects must be clearly understood.
● Current performance must be accurately monitored to ensure that any changes will have the desired effect. For example, increased emphasis upon piglet mortality will be of little use if litter size remains too low.

2. How can better ways of operating the unit be discovered?

● Simple organisational problems can normally be solved by a general discussion among members of the team.
● The pigman may read details on the problem he is seeking to resolve.
● He may attend meetings, or courses.
● Expert specialist advice may be sought.
● No one person is ever in possession of all the information on a subject and new technology may be available but unknown to the pigman—thus talking to others is a prime source of new information.

3. How can the best course of action be decided?

● This requires a particular and not common ability to weigh up what is, particularly with livestock, often conflicting advice. However, it may help to:
(a) consider the effects of each option on staff;
(b) consider the owner's need to invest more money;
(c) assess the time available to carry out any change;
(d) determine the likely benefits of the change—i.e., is it really worth while to alter the routines?

4. How can others be persuaded to change?

● Prepare details for a discussion with those who will be involved, listing:
(a) the problem and need for change;

(b) the options and how each may affect the unit, the results and the operators.
- Make it clear that everyone involved should state their own views.
- Time meetings so that people are relaxed. They should not be anxious about work they have to do or about leaving for home. Good timing will emphasise how important is the need for change.
- Give the quieter members of the team every encouragement to express their view—a greater commitment to change will arise if everyone feels that they have been given a chance to contribute.
- Do not worry if people contradict themselves when talking—this often means that they are changing their view.
- Try to get a consensus—preferably unanimous—agreement on a new course of action and give credit to those who contribute opinions when summarising.

5. *How will everyone remember the new routine?*
- Proper emphasis will be placed upon any change if an action plan is drawn up indicating:
  (a) the new plan;
  (b) who is involved;
  (c) what results are expected.

6. *How can the momentum from a change be maintained?*
- It is vital that all unit actions are monitored. With a change in routine everyone will achieve greater credibility if the effects of a change are reviewed after a period of time.
- Regular checks on the changes will make it:
  (a) easier to influence others in the future;
  (b) more likely that colleagues will think more deeply about the way that they carry out their work.

Remember
This style of leadership stems from understanding the needs of the owner as well as the aspirations and career needs of others. In our society coercion is not an acceptable form of change and consultation will invariably yield a better and longer-lasting result. Because the pigman has such a major effect on performance his opinion and co-operation simply cannot be ignored.

Chapter 21

The pigman has to understand pigs—and people. He does not need to be trained in business studies or high finance.

However, he ought to appreciate something of the problems of those who finance the pig business. He might also benefit by understanding some of the terms used to describe the financial measurements of efficiency. This will assist him to appreciate the relationships between certain physical and financial characteristics.

In this chapter some basic preparation for more senior management responsibilities will also be considered.

Any investment must be made with a view to the successful increase in the sum invested. For example, at a personal level a pigman may consider a savings account with the Post Office or a Building Society. As a private 'investor' he is interested in the rate of interest, security and access to his money for sudden unforeseen needs.

The farmer or businessman investing in pigs has a parallel set of considerations. Firstly, he considers the anticipated increase in the value of his money. The more successful the pig unit the faster and larger his money grows—this equates to the saver's 'rate of interest'. As for security, the pig unit investor wants to make sure that he will not go broke and lose his money. So he makes judgements which are based on the likely levels of costs and returns in the future, levels of output achieved, building costs and so on. The business investor has reduced access to his money and can only 'realise' his assets by disposing of the business or using the pig unit as security against which he may borrow spendable ('liquid') money.

Thus, the pigman concerned for his own savings and his future financial security, faces on a different scale the same basic considerations as the proprietor of the pig business. There is another basic difference, of course. The pigman has no one but himself to account for in the disposition of his personal savings. The owner of the pig unit relies heavily upon his staff for a commitment and level of return on his investment, highlighting the vital role played by the pigman.

There is a strong relationship between the amount of money

invested in buildings and equipment and the number of pigs tended per man. For example, on a unit where feed is purchased in bags, dispensed by hand and bedding and manure moved by staff the labour requirements (but *not* initial cash requirements), will be higher than where bulk bins, automatic feeders and automatic manure removal have been provided by a higher financial outlay.

Further there is a point, as illustrated by certain national recording schemes, below which unit performance tends to fall off if investment in labour is at too low a level.

Superimposed upon these considerations is the truly massive relationship between the degree of effective operator input and profit.

The pigman will hear expressions used in discussions on unit finances from time to time. The shortlist on page 233 gives the normal definitions associated with the terms used.

The pigman has no influence over the cost of buildings and equipment. It should be realised, though, that where labour-saving fittings are used, it is necessary to produce more pigs per man. Thus it is impossible to compare the number of sows per man on units with different levels of fixed capital investment. At any stage when the owner installs modern equipment or appliances he will expect to see unit efficiency increase—less pigs lost, quicker growth, an expansion with no extra staff are examples. It is quite naïve for a pigman to assume that money can be invested in a pig unit just to make life easier for him. The owner will expect 'easier for the staff' to be mirrored by 'so we will get better results'.

Where the pigman has an influence is in helping by his efficiency to achieve a higher output level. This will reduce the amount of money borrowed, thus reducing the interest bill; and it will also encourage the owner to invest more in the unit.

Attention to the vital aspects of husbandry outlined in the preceding chapters of this book will reduce the expenditure on the day-to-day requirements of the business. For example, careful setting of a weaner house control panel may not just ensure saving of expensive fuels, but may also encourage piglets to consume more feed and gain more efficiently so that they spend less time on the unit.

When financial measures of efficiency are considered it has to be borne in mind that the pigman may not be in a position to influence the price of the goods coming on to the unit. However, as stressed, he does play a major part in deciding upon the degree to which they are efficiently used. Careful usage and storage as well as a high level of output are continuously within the grasp of the operator.

Terms used	Accepted definition	Explanation
Capital	Monies required to establish and operate the business.	The owner's financial outlay.
Fixed Capital	Money used to purchase buildings, equipment and instal services, (drains, roads, water, etc.) and normally includes initial cost of the stock.	This is the investment needed to start the business or change or renew the unit fabric.
Working Capital	Sum required once the need for fixed capital has been satisfied. Normally includes feed, labour, veterinary costs, bedding, repairs, etc.	These are the 'household' bills. Pigmen can influence these greatly by their efficiency.
Cash Flow	Prediction of financial needs or surpluses against a time span. May include fixed and working capital.	Shows those periods when the total business should be in the 'black' or in the 'red'.
Gross Margin	The financial status of the unit based upon working capital surplus/deficit. Usually expressed on a 'per sow' or 'per pig sold' basis.	Allows a quick check of unit efficiency over major running costs.
Margin Over Feed	The difference between income and feed costs.	Helps to check the stockman's efficiency.
Output	The income from sales.	The money received—like wages before stoppages.
Return on Capital Invested	Once all costs of operating the business have been deducted from output, the surplus can be judged like an interest rate to show if the Fixed Capital has been wisely invested.	This can be said to measure the suitability of the unit design and the efficiency of the pigman in achieving economic output.
Finance Charges	The cost of borrowing money for the establishment and operation of the business.	Same as bank overdraft or hire purchase charges.

It is a vital necessity for each and every pigman to understand the close relationship between relatively small improvements in efficiency and profit. The classic example is in sow productivity which is shown below, but other equally important examples could be used.

In this example, the income from each weaner is taken as £26.25 and the influence upon returns, in this case, of an increase in sow productivity can be seen.

Share of depreciated building cost per litter produced: £14
Cost of feed used per litter: £132
Share of other working capital costs per litter: £22

   TOTAL COST of producing one litter =   £168

Thus, for a sow producing 9.6 pigs per litter sold, the gross income is (9.6 × £26.25 sale price per weaner) = £252 per litter
Therefore, if 9.6 weaners per litter are sold the margin, before finance charges, is: (£252 − £168) = £84 per litter

If, however, efforts are made to increase the output per sow to 9.8 pigs sold per litter, the gross income improves to (9.8 × £26.25 sale price per weaner) = £257 per litter
This gives a margin of (£257 − £168) = £89 per litter.

If sows produced 2.35 litters per year, in this example £5.00 increase in margin per litter leads to an increased income *per sow* of £11.75. So for every 100 sows, the additional margin resulting from the increased effort by the pigman would be £1,175.

Remember

This simple example illustrates the difference in profitability of a sow enterprise which may arise from a small (in this case a 2 per cent) improvement in efficiency. It is especially important to realise that the pigman can almost always affect output by *at least* this amount and usually by far more.

Chapter 22

It has been a long-established feature of employer/employee relationships to reward on results. Piecework and bonuses are accepted features of British agriculture.

Compared to workers in many other industries, the pigman has to work unsocial hours. Frequently the work that has to be undertaken includes a proportion of unpleasant yet essential tasks. Further, the degree of responsibility and day-to-day decision-making is frequently greater than in many other industries with comparable rates of pay.

However, it is a fact that statutory agricultural wages fall below those of many other skilled and semi-skilled trades. Many pigmen enjoy wages which compare favourably with other trades because their employers acknowledge their important contribution to the success of the business. In particular, in agriculture there are more workers living in dwellings owned by their employers than in almost any other industry.

The methods of reward for effort and skill are numerous. Experience has shown, however, that piecework and bonus payments are much less suited to livestock production than to arable or market gardening enterprises. This is because it is harder to predict all influences upon the productivity of a livestock enterprise and also because increased pig output invariably has an additional cost element associated with it—unlike cutting a field of cauliflowers, for example.

Is there any one best way of payment for the efforts of a pigman? The answer is probably—no!

Eliminating bonus payments and the like means that misunderstandings and misinterpretation can be avoided because the regular review process provides adequate opportunity for the employer to reward his staff for improvements in efficiency or to share with them the fruits of a favourable spell of trading.

Some employers prefer to agree an annual or periodic bonus payment with their staff or key members of the team. This might be workable given a very clearly defined interpretation of the point at which 'profit' or 'margin' is calculated and upon which the bonus payment is made.

The difficulty with bonus or profit-sharing comes in years where pig prices do not adequately match costs of production. This is frequently outside the control of the pigman and his boss. It is at this point that both employer and employee become dissatisfied and where a pigman who has made a particular effort to improve his efficiency finds that no reward is forthcoming. It may, therefore, be much fairer if the pigman is rewarded for his efforts without taking into account the vagaries of international grain prices, the demand for pigmeat or other factors beyond his control.

In an attempt to avoid the problems caused by outside factors, various efforts have been made to calculate incentives based upon certain elements of physical performance. Quite frequently 'head-age' payments for numbers of pigs sold or weaned have been used as a basis for such schemes. Sometimes these payments are a flat rate based on the total of pigs produced; in other cases the unit payment increases once a predetermined number of pigs has been exceeded.

Once again, any such scheme calculated on this basis—or any other yardstick—is open to a degree of manipulation or interpretation which may not be in the best interest of either party. The pigman might feel aggrieved at a higher than normal influx of gilts in the breeding herd which reduces herd output in a given period. Equally, an employer might view feed usage to gain higher physical output unjustified. Dissatisfaction breeds distrust and this inevitably leads to a breakdown in the good human relations so vital to a profitable unit.

Thus, it is probably more satisfactory to negotiate a set rate for a job well done and to encourage a free and frank exchange of views about unit performance and terms of employment and remuneration at regular intervals.

Turning from cash considerations and the role of financial incentives, just what method can be used to motivate staff other than through money-in-the-pocket?

The first thing is money-in-the-pocket—No that is not a misprint! As stated previously, a reasonable reward for the efforts made—reasonable to employer and recipient—is essential for the proper motivation of any pigman at any level. However, as stated at the end of Chapter 19, this is probably less important, in the final analysis, than satisfaction in the job, having a sense of belonging to a successful team and being recognised for the part that even young trainee pigmen can play in that success.

Now this may not sound a tall order but experience shows that it can be an insurmountable one under some circumstances. Why is it so difficult to provide job satisfaction and, more important, what might be done to improve this?

Aside from difficulties in private life, which have been referred to in previous chapters, the main area of concern is friction between individuals—often with faults attributable to both, or all sides. This has to be tackled quickly and a strong effort made to ensure that grievances are settled once and for all. The complexity of the human nature makes that difficult, but tolerance and better communications are the cornerstones of harmonious relationships as considered at length in Chapters 19 and 20.

Other de-motivating sources can be traced to working conditions. A lot of menial, physical tasks dampen enthusiasm for pig care. It is, after all, difficult to remember that pig output is the main aim when confronted with a yard of muck and a fork plus wheelbarrow. Conversely, much can be done to reorganise routines so that the less pleasant and more arduous physical tasks are shared among more people, or so that these jobs can be carried out with less physical 'slog'. Two simple improvements are delivering goods and feed to a convenient point nearer to pigs, and making a simple board mounted on a stand to simplify pig movement. Therefore, instead of complaining about the poor layout of the system, the pigman should first stand back and think what can be easily done to improve the job and make it easier.

Such an approach is hardly ever wasted. Sometimes, we can think of no way of improving a routine ourselves and have to rely upon others to help—so be it. At least the job may be made easier so that the operator would feel his burden lighten and become more efficient and better motivated.

It is not always possible to improve matters satisfactorily by simply undertaking routines another way. That does not mean that the pigman gives up or becomes disenchanted. He, or she, should prepare a case, and then follow the suggestions made in Chapter 20 as to how to present it at an appropriate time to the more senior worker or owner.

Another quite common side to the question of satisfaction is where the pigman feels frustrated by a lack of knowledge or experience to deal with a problem that he can identify. Reference has been made to the various sources of information, but these will not always reveal the vital point which will help the operator. It must be hoped that both pigman and employer recognise this need together, and plan to correct it.

By far the most effective training is conducted on the farm itself, so that full weight can be given to the context of the task, but there are occasions where the Agricultural Training Board or Local

Education Authority can be used to tailor instructional courses to uprate technical ability and knowledge. There is a gradual change in training to the operation of training sessions to suit a need, rather than a broader-based, more generalised course. Thus, if the pigman and his owner can identify the need for training and this need cannot be appropriately dealt with on the farm, the local office of the Training Board may be approached. There are probably other operators in the area with similar problems and a course can often soon be established and run.

The involvement of the pigman in the operation of a unit has already been stressed as an important part of improving job satisfaction. The pigman must not presume that his ideas will always prevail, but he should be ready to help others in the team with suggestions rather than criticisms.

Both owner and pigman should be willing to attend meetings or events where they may together consider other methods. Such semi-social occasions provide a relaxed opportunity for everyone to discuss their opinions and problems. As well as gleaning ideas from other sources the owner and pigman will be able to talk over matters which have caused concern on their own farm but without causing real problems of personality.

By far the most satisfactory introduction for a new employee is a probationary term of employment of 3–6 months followed by a review by both parties at the end of this period. This allows each to assess the capabilities and character of the other. The employee can also sum up the degree of effort that the unit design and structure require of him, and both parties can then settle on a payment package. Annual reviews are also sensible ideas, so that Annual Agricultural Wages Board conditions can be incorporated and any other adjustments to changed conditions or responsibilities made. If employee and employer alike agree to such a plan, it is normally possible for a single weekly or monthly payment schedule to be agreed without recourse to any other incentives.

Finally, we all like to feel appreciated. Just because the employer or his agent does not tell the pigman that he is pleased with his contribution, does not mean that the pigman should not tell those junior to himself of *his* satisfaction! There is much to be gained from an approach which allows credit to be passed on when due—after all it provides a route for passing on criticism when that, too, is appropriate!

Thus, the pigman, at all levels of seniority, can achieve satisfaction by:

- giving thought to what modifications he can make to improve working conditions and routines;
- being willing to attend functions and seek opinions about improving routines;
- admitting when insufficient knowledge or skill is limiting achievement in the job;
- readily helping others in the team with their problems and helping to contribute to possible remedies;
- receiving praise humbly and criticism constructively, and handing out, in his own turn, both in a similar manner.

A final form of motivation is to be given the opportunity for advancement, or promotion. Every pigman must realise that there are more people who consider themselves suitable for promotion than there are opportunities. Therefore, patience must be exercised and an understanding that it takes more than good stockmanship to make a person suitable for leadership.

Reference Glossary

Normal body temperature of a pig 39-39.5°C (102-103°F)
Pulse rate 70-80 per minute in adult pigs at rest but up to 250 per minute in baby pigs; 20-30 per minute in finishing pigs at rest (lower in adult stock but higher in younger pigs and where exposed to higher environmental temperatures).
The mature bodyweight of a white hybrid sow will be around 200-220kg (440-484lb)
The preferred weight of a gilt at first mating is 120-130kg (264-286lb)
The preferred age of a gilt at first mating is 33-35 weeks
A sow will eat (including boar's share and maiden gilt's share) 1.1-1.25 tonnes/year indoors; 1.4-1.5 tonnes/year outdoors
Piglets under one week of age require a lying zone temperature of 32° (90°F)
Pigs under one month of age require a constant air temperature no less than 27°C (80°F)
For every 100 sows in a herd there will be a need to replace 35 per year
Pigs should eat the following average daily quantities:

- At 4 weeks of age   350g/day   ($\frac{3}{4}$lb/day)
- At 8 weeks of age   900g/day   (2lb/day)
- At 12 weeks of age   1.35kg/day   (3lb/day)
- At 16 weeks of age   1.90kg/day   ($4\frac{1}{4}$lb/day)
- At 20 weeks of age   2.60kg/day   ($5\frac{3}{4}$lb/day)

1kg (2.205lb) live weight = approximately 750g ($1\frac{3}{4}$lb) carcase weight.
1kg change in carcase weight = approximately ± 0.2mm in P_2 backfat
1 day change in age = approximately ±1mm in carcase length
Boars may average 2.5mm greater carcase length at the same dead weight as gilts, which are, in turn, a similar amount longer than castrates.

Index

241

FARMING PRESS BOOKS

Below is a sample of the wide range of agricultural and veterinary books published by Farming Press. For more information or for a free illustrated book list please contact:

Farming Press Books, Wharfedale Road, Ipswich IP1 4LG, Suffolk, Great Britain.

Housing the Pig
Gerry Brent

Provides guidelines to assess proposals for investment in buildings and equipment and includes fifty detailed layouts for all classes of stock, integrated systems and ancillary services.

The Growing and Finishing Pig: Improving its Efficiency
P. R. English, S. H. Baxter, V. R. Fowler and W. J. Smith

A large, comprehensive volume which explores in detail the factors that control the efficiency of the pig from weaning to slaughter.

Practical Pig Production
Keith Thornton

An excellent introduction to pig farming, full of practical advice.

Pig Ailments
Eddie Straiton

This 6th edition of the TV Vet Book for Pig Farmers is an outstanding pictorial guide to pig diseases giving clear details about treatment and prevention.

Pig Diseases
David Taylor

The fifth edition of a technical reference book based on teaching notes, for the veterinary surgeon and pig unit manager. (Published by the author)

Pigmania
Emil van Beest

A collection of cartoons presenting the lighter side of porcine activity.

The Economics of Pig Production

A detailed survey of pig production from 1946-91. Includes analysis of stock and management systems, unit production and efficiency.

Any Fool Can Be a Pig Farmer
James Robertson

A classic account of an innocent stumbling into pig-farming

Farm Building Construction
Maurice Barnes and Clive Mander

Practical information covering all aspects of farm building, from initial planning onwards.

Farm Machinery
Brian Bell

Gives a sound introduction to a wide range of tractors and farm equipment. Third edition enlarged and incorporating over 150 photographs.

Farming Press also publish three monthly magazines: Dairy Farmer, Pig Farming and Arable Farming. For a specimen copy of any of these magazines please contact Farming Press at the above address.